Development of continuous processes

for aerobic glycerol oxidation

Zur Erlangung des akademischen Grades eines

Dr.-Ing.

von der Fakultät Bio- und Chemieingenieurwesen
der Technischen Universität Dortmund

Dissertation

vorgelegt von

Ken Aulia Irawadi, M.Sc.

aus

Bogor, Indonesien

Tag der mündlichen Prüfung: 27.05.2013

1. Gutachter: Prof. Dr. rer. nat. Arno Behr

2. Gutachter: Prof. Dr.-Ing. Rolf Wichmann

Dortmund 2014

Bibliographic information published by the Deutsche Nationalbibliothek

The Deutsche Nationalbibliothek lists this publication in the Deutsche
Nationalbibliografie; detailed bibliographic data are available
in the Internet at http://dnb.d-nb.de .

ISBN 978-3-8325-3642-8

Logos Verlag Berlin GmbH
Comeniushof, Gubener Str. 47,
10243 Berlin
Tel.: +49 (0)30 42 85 10 90
Fax: +49 (0)30 42 85 10 92
INTERNET: http://www.logos-verlag.de

ABSTRACT

Converting glycerol into a more valuable substance is especially important if the glycerol price declines due to a oversupply from the biodiesel industry. Aerobic catalytic glycerol oxidation is meant to improve the commercial value of glycerol through an environmental friendly method.

Previous works in this field carried out batch-wise had promoted the utilization of heterogeneous catalysts of palladium, platinum and gold supported on materials such as activated carbon, TiO_2, and Al_2O_3. This dissertation elaborates the development of a continuous catalytic glycerol oxidation.

The work was started by a reexamination of commercially available palladium and platinum catalysts. Both metals were supported on Al_2O_3 and activated carbon pellets. By using the palladium catalyst it was possible to produce batch-wise glyceric acid with a maximum yield of 34 %, while generating tartronic acid as another product with a yield of 8 %.

When the same reaction conditions were applied in a continuous mode, a declining activity was observed. Although the activity could be restored by reducing the catalysts at 300°C in hydrogen atmosphere, this treatment will interrupt the continuous process. Therefore, it was decided to develop a slurry system where the catalysts reactivation can be carried out ex-situ.

After comparing the settling speed of different potential supports such as titanium, aluminum and iron oxide powders it was found that the magnetic particle Fe_3O_4 is suitable to be used in the slurry system. This was followed by the development of a miniplant which consists of a continuous stirred tank reactor (CSTR) and electromagnetic separators. Catalyst gold-platinum on magnetic Fe_3O_4 particles was used in the miniplant. The reaction ran continuously and the catalyst was magnetically separated and reused for the next reaction. Inductively coupled plasma (ICP) analyses found no metal catalyst traces in the residue (< 1 ppm). The yield towards glyceric acid was 26 %.

ZUSAMMENFASSUNG

Die Umwandlung von Glycerin in höherwertige Verbindungen ist insbesondere im Hinblick auf den durch ein Überangebot aus der Biodieselindustrie verursachten Preisverfall von großem Interesse. Die aerobe, katalytische Oxidation von Glycerin soll als umweltfreundliche Methode den kommerziellen Wert von Glycerin steigern.

In vorhergehenden Studien haben sich für diskontinuierliche Prozesse heterogene Katalysatoren wie Palladium, Platin und Gold auf Trägermaterialien wie Aktivkohle, TiO_2 und Al_2O_3 als günstig erwiesen. In dieser Dissertation wurde ein Verfahren zur kontinuierlichen, katalytischen Glycerinoxidation entwickelt und ausgearbeitet.

Die Arbeit wurde zunächst mit einer erneuten Prüfung der kommerziell erhältlichen Palladium- und Platinkatalysatoren begonnen. Beide Metalle wurden mit Al_2O_3 und Aktivkohlepellets als Trägermaterialien eingesetzt. Bei Verwendung eines Palladiumkatalysators konnte unter diskontinuierlichen Bedingungen für Glycerinsäure eine maximale Ausbeute von 34 % erreicht werden, für Tartronsäure als weiteres Oxidationsprodukt eine Ausbeute von 8 %.

Bei Anwendung der gleichen Reaktionsbedingungen in einem kontinuierlichen System wurde eine Abnahme der Aktivität beobachtet. Obwohl die Aktivität des Katalysators durch Reduktion bei 300°C unter Wasserstoffatmosphäre wiederhergestellt werden konnte, unterbricht dieser Schritt doch den kontinuierlichen Prozess. Daher wurde ein Slurry-Reaktor entwickelt, in dem die Reaktivierung des Katalysators ex situ durchgeführt werden kann.

Nachdem die Sinkgeschwindigkeit verschiedener, potenzieller Trägermaterialien wie Titan, Aluminium und Eisenoxidpulver untersucht wurden, stellten sich magnetische Fe_3O_4-Partikel als geeignet für den Slurry-Reaktor heraus. Im nächsten Schritt wurde eine Miniplant bestehend aus einem Rührkessel (continuous stirred tank reactor, CSTR) und elektromagnetischen Separatoren entwickelt. In der Miniplant wurde ein Gold-Platinkatalysator auf magnetischen Fe_3O_4-Partikeln verwendet. Die Reaktion wurde kontinuierlich betrieben und der Katalysator magnetisch abgetrennt und für die nächste Reaktion wiederverwendet. Mittels ICP-Analyse (inductively coupled plasma) konnten keine Spuren des Metallkatalysators (< 1 ppm) im Rückstand nachgewiesen werden. Die Ausbeute für Glycerinsäure war 26 %.

ACKNOWLEDGEMENTS

I would like to express my sincere gratitude to my advisor Prof. Dr. rer.nat. Arno Behr for giving me the opportunity and valuable support to carry out my PhD work in his chair. I am also very grateful for his scientific advice and helpful suggestions. My thanks are also delivered to Prof. Dr.-Ing. Rolf Wichmann for being the second assessor of my PhD examination. His inputs contributed to the perfection of my dissertation. I also thank Prof. Dr.-Ing. Andrzej Górak for being the examiner of my PhD examination.

To my former team members, Abderrahim Oussaa, Abdo Aldeiri, Ahn-Vu Lee, Antonia Neumann, Cai Yu, Friederike Wrobel, Ina Schönfeld, Nga Tran, Pierre Schwach, and Tim Seifert, I owe them my heartfelt thanks.

I wish also to thank my friends, Dr. Alaeldin Bouaswaig, Dr. Babu Halan, Dr. Falk Lindner, Dr. Ganjendran Kadasamy, Dr. Guido Henze, Jens Eilting, Dr. Julia Leschinski, Dr. Marc Becker, Dr. Marc Dumont, Dr. Mattijs Julsing, Nandian Syarif, Dr. Rohan Karande, Dr. Stephan Dech, Dr. Thomas Beckmann, Dr. Ulf Schüller, and Dr. Yudy Tan, for all their help and insightful discussions.

I would like to thank Prof. Hans-Peter Niedermeier, Dr. Michael Czepalla and Mr. Christian J. Hegemer from the Hanns Seidel Foundation for all generous support. I am very grateful to Emery Olechemicals for the financial and material support. My thanks are also delivered to Dr. Fuadi Rasyid from the Habibie Center. Moreover, I would like to thank Prof. Dr. Jörg Tiller and Ms. Monika Meuris from the chair of Biomaterials and Polymers Science as well as Prof. Dr. Heinz Rehage from the chair of Physical Chemistry at TU Dortmund for their generous help with analytics.

Finally, I would like to thank my family, my mother Prof. Dr. Tun Tedja Irawadi, my father Prof. Dr. Irawadi Djamaran and my wife Kathrin Botsch-Irawadi, To me, their patient support and encouragement have meant more than I could express.

SYMBOLS AND ABBREVIATIONS

Symbol	Unit	Name
c	[mol/l]	Concentration
c_G	[mol/l]	Concentration of gas
K_H	[m^3 bar mol^{-1}]	Henry's coefficients
M	[mol/l]	Molarity
n	[mol]	Molar amount
p_G	[bar]	Partial gas pressure
S	[%]	Selectivity
T	[°C]	Temperature
t	[h]	Time
τ	[h]	Retention time
X	[%]	Conversion
Y	[%]	Yield

Abbreviation	Meaning
AC	Activated Carbon
DHA	Dihydroxyacetone
DLS	Dynamic Light Scattering
EDS	Energy-Dispersive x-ray Spectroscopy
ee	Enantioselectivity
FTIR	Fourier Transform Infrared Spectroscopy

Abbreviation	Meaning
GLA	Glyceric Acid
GLD	Glyceraldehyde
ICP	Inductively Coupled Plasma
Mt	Megaton
MCM	Mobil Composition of Matter
NP	Nanoparticles
n.d.	Not determined
PVA	Polyvinyl Alcohol
SEM	Scanning Electron Microscope
TEM	Transmission Electron Microscope
TMC	Transition Metal Catalysts
TOF	Turnover Frequency $[h^{-1}]$
TON	Turnover Number
UV-VIS	Ultraviolet Visible Spectroscopy
XPS	X-ray Photoelectron Spectroscopy
XRD	X-ray Diffraction

TABLE OF CONTENTS

1 INTRODUCTION

1.1 Background

As a byproduct of oleochemical industries, glycerol is produced proportionally with the production of oil and fat derivatives. Today, the largest contributor to glycerol production is the biodiesel industry, where about 100 kg of glycerol is generated for every ton of biodiesel produced. This accounted for about 2.7 million tons of glycerol produced worldwide in 2012 from the biodiesel industry alone.[1] In the long run, due to environmental regulations in major industrial and developed countries, such as the EU directive 2009/28 that obligates the utilization of renewable energies to fill 10 percent of transportation energy demand by 2020, the global biodiesel production will increase severalfold.[2]

It is expected that the biodiesel production will grow worldwide from 27 million tons in 2012 to 35 million tons in 2017, respectively.[1] This condition may soon lead to a huge surplus in the glycerol market and eventually will decrease its economic value. Moreover, the absence of any particular action that deals with this issue will not only give an economic disadvantage to oleochemical companies but also cause serious environmental problems.

One possible way to solve this future problem is to expand the field of glycerol utilization. Along its conventional use in cosmetic, pharmaceutical or food industries, glycerol might also have a completely new function, for example as a growth medium for oil-producing microalgae or as a feedstock for poultry[3]. Alternatively, a more effective way may be achieved by chemically converting glycerol into more economically valuable products such as dihydroxyacetone (a main compound in sunless tanning agents), glycerol tertiary butyl ester (a potential octane booster) or propanediol (a major important basic chemical).[4]

1.2 Motivation

Chemical conversion of glycerol can be performed in a more efficient and nature friendly way when carried out catalytically. Especially for oxidation reactions, the use of catalysts may refrain industry from utilizing non environmental friendly oxidants such as chromate based compounds. For glycerol oxidation reactions, environmental friendly oxidants such as air or oxygen gas can be used with the presence of heterogeneous palladium[5], platinum[6] or gold[7] metal catalysts, thereby creating a positively significant effect to the environment.

The reactions catalyzed by heterogeneous catalysts usually take place on the surfaces or the pore surfaces of the respective catalysts. Hence, decreasing the catalyst particles size will increase the area for the reaction. Moreover, a metal element cluster sized less than 100 nm might have a significantly different behavior to a bulky form of the same element. Therefore, attempts are made to reduce the catalysts size, for example by introducing nanoparticle technologies to their synthesis processes. Since these technologies enable the production of very small metal particles, i.e. with the size in the range of 0.5 to 100 nm, the progress of catalytic glycerol oxidation reaction may be influenced significantly by the development of these technologies.

A catalytic reaction at an industrial scale is usually realized as a process that runs continuously. There are at least three types of reactor systems based on the way heterogeneous catalysts are introduced into a reactor of a continuous process, i.e. fixed bed, fluidized bed and slurry systems. For the first two systems, the catalysts are retained inside the reactor during and after the reaction. With these systems, no additional catalysts separation step from other reactants is required. However, if the catalysts deactivate rapidly during the reaction, the continuous process should be interrupted too often to reactivate the catalysts. On the contrary, in the slurry system, the catalysts flow together with the reactants and products into and out of the reactor, respectively. While this system requires an additional separation step, nevertheless, it offers a possibility to reactivate the catalysts ex-situ; thus, no interruption is needed during the reaction.

The separation of very small catalyst particles, especially in a nanometer scale, via common processes such as filtration or sedimentation is hardly practical at an industrial scale. Therefore, the application of nanosize catalysts will only be possible if a new catalyst separation method suitable for a continuous industrial scale process is available. Fortunately, magnetic fields have been utilized as a means of small metal particle separations, especially in water treatment plants. Therefore, it is expected that the application of magnetic field can answer this separation challenge.

1.3   Thesis objectives

The main purpose of this dissertation is to seek the suitable methods for conducting an industrial scale glycerol oxidation process. This main idea is divided into specific objectives. The first objective is to **examine the performance of commercially available heterogeneous catalysts** for this process. Since thorough investigations of the batch-wise reaction on the laboratory scale have been conducted elsewhere, in this dissertation only the most important parameters will be reevaluated, for example, the optimal reaction temperature and pressure as well as the concentration of reactants. Based on the reexamined parameters, in a miniplant-scale fixed bed reactor system, continuous glycerol oxidation reaction will be carried out. Observing the performance of the system, the characteristic traits of the industrial scale process may be predicted accurately.

Due to their high selectivity, the utilization of homogeneous catalysts may improve the performance of glycerol oxidation process. However, considering their challenging separation from the reaction medium after the reaction, their application in an immobilized form is rather mandatory for this process. Therefore, the second objective is to **evaluate an immobilization method of a homogeneous catalyst**.

Based on the results of the continuous fixed bed reactor experiments, the third objective of the dissertation is to **develop a system that enables the utilization of a slurry-formed catalyst**. In this system, the catalyst particles between 100-300 nm, together with the reactants, will enter the reactor where the glycerol oxidation reaction takes place. They will leave together with the products and the rest of the medium. In this system, an additional step to remove the catalysts from the products is necessary. Since only magnetic catalyst particles that can be used in this system and this type of catalyst are not commercially available yet, the fourth objective of this dissertation is to **develop the magnetic catalysts** for the catalytic glycerol oxidation slurry system.

Along its role as a raw material for the production of other commercially important intermediate substances, glycerol could also play a role in the synthesis of metal nanoparticles. This is especially possible since glycerol has a potential as a reducing agent which is required for the production of metal nanoparticles. The success in this field will facilitate the substitution of non environmental friendly chemicals that are usually used in this process. Therefore, the fifth objective of this dissertation is to investigate **the utilization of glycerol in the production of metal nanoparticles**.

1.4   Dissertation organization

The second chapter of this dissertation will review important information and previous research relevant to this topic. In section **2.1** of this dissertation, more detailed information about glycerol production and market as well as the derivative products from glycerol oxidation reaction will be described. The development of heterogeneous catalytic oxidation systems for glycerol and other alcohols will be covered in sections **2.2** and **2.3**. In section **2.4**, several synthesis methods of catalyst metal nanoparticles will be described. These methods are also useful for the production of magnetic nanoparticles. Furthermore, section **2.5** will introduce the basic concept of magnetic separation. In addition, section **2.6** will introduce different type of reactors for continuous catalytic reactions.

In the third chapter, the result of the experiments will be presented and discussed. The performance of commercial catalysts for glycerol oxidation reactions will be presented in section **3.1**, whereas in section **3.2** an immobilization method of a homogeneous catalyst is evaluated. Section **3.3** will focus on the development of the reactor system for the slurry catalytic oxidation process. In section **3.4**, the development of magnetic slurry catalysts for the glycerol oxidation process will be described, whereas its financial aspect will be appraised in section **3.5**. Moreover, section **3.6** will cover the utilization of glycerol in the synthesis of metal nanoparticles. Finally, the conclusion and the experimental part of the dissertation will be presented in chapter **4** and **5**, respectively.

2 GENERAL PART

2.1 Glycerol as a renewable raw material in chemical industry

Glycerol, 1,2,3-propanetriol, is the simplest triol found in all natural fats and oils as fatty esters. Glycerol can also be obtained chemically from propene or via fermentation of sugar. Today, the fastest growing sources of glycerol come mainly from oleochemical and biodiesel industries as a byproduct in the conversion of fats and oils to produce fatty acids or fatty acid methyl esters. The process is carried out either through saponification (yielding glycerol and soap), hydrolysis (yielding glycerol and fatty acid), or transesterification (yielding glycerol and fatty acid methyl ester).[4] About 10% (by volume) of glycerol is produced as a by-product in these processes.[8]

The industrial application of glycerol started in early 1866 when nitro glycerol, a main component of dynamite was produced. Today, glycerol has numerous industrial applications such as a conditioner and a moisturiser in food and cosmetic industries, as an emulsifier in mono- and diglycerides form, as a mild laxative in pharmaceutical applications, as a plasticizer in cellulose films and as a moisturizer in tobacco, etc.[8] Figure **2.1.1** shows various fields in which glycerol is used (given in volume %).

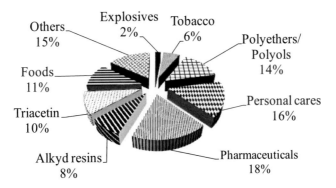

Figure 2.1.1: Industrial application of glycerol (volume %).[9]

2.1.1 Glycerol market

The world's biodiesel production has increased from around 912,000 m^3 in 2001 to more than 12 million m^3 in 2008. Moreover, it was expected to grow to 23 million m^3 by the end 2013.[10] In Europe, the growth of biodiesel was strongly encouraged by the European Union's ambition in regenerative energy, such as the EU directive 2003/30/EC which instructed all member states to substitute 5.75% of petroleum fuels with bio-fuels for by 2010.[11]

The growth of biodiesel production increases the supply of glycerol worldwide. For example, in 2007 biodiesel manufacturers produced 187,000 tons of crude glycerol and other 150,000 tons of crude glycerol were produced through other processes.[12] As the result of oversupply, the glycerol price steeply decreased in the market.

However, when discussing the price of glycerol, it is important to differentiate the market price for crude glycerol and for refined glycerol. While the price of crude glycerol depends on the amount of its feedstock, i.e. the volume of fatty acids and fatty acid methyl esters (e.g. biodiesel) productions; the price of refined glycerol depends strongly on the capacity of glycerol refineries, instead of the amount of crude glycerol feedstock.

By observing the price movement of refined glycerol, there was a time when the price had declined significantly. For example, between 1995 to 2005, the prices of refined glycerol in Europe and USA had decreased to 50%.[13] However, when we observe the price of refined glycerol in a shorter time frame, there were periods when the price is significantly high. For example, as showed in figure **2.1.2**, in the beginning until middle of 2008 the price of refined glycerol was varying between 800 and 1000 €/ton, while at the end of the year the price fell down to 50%.[14] Recently, in Asia the price is relatively stable (see figure **2.1.2**).

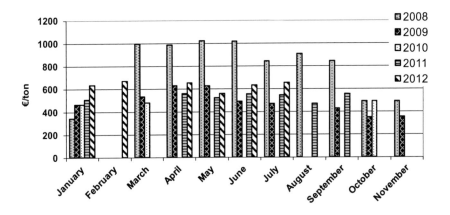

Figure 2.1.2: Price of refined glycerol in Shanghai market from 2008 to 2012.[14]

2.1.2   Products from catalytic glycerol oxidation reactions

To stabilize the fluctuating tendency of glycerol price, new alternatives to utilize glycerol should be investigated. However, since the direct utilizations of glycerol have been already intensively explored, it seems the research in this field should be more focused on the production and utilization of glycerol's derivative products. Comprehensive examples of glycerol's derivatives produced catalytically have been presented with the review of Behr et al., for example, monoglycerides (esterification products of glycerol with carboxylic acids), glycerol tertiary butyl ether or GTBE (etherification product of glycerol with isobutene or tert-BuOH), 1,2-propandiol (via hydrogenation of glycerol), glyceric acid (glycerol oxidation product), etc.[4]

Glycerol oxidation can deliver a wide range of products due to its complex reaction pathway (see figure **2.2.3** and appendix **A5**). Several valuable products from glycerol oxidation reactions are discussed as follows:

a. **Dihydroxyacetone** (DHA) is produced when the secondary alcohol of glycerol is selectively oxidized. Conventionally, it can be produced from glycerol by a fermentation process, e.g., using *Gluconobacter oxydans*.[15] It is used mainly as an active substance in sunless tanning lotions. To improve the performance of dihydroxyacetone, it can be incorporated with different emulsifiers, for example phospholipids.[16]

b. **Glyceraldehyde** (GLD) is a product of glycerol's primary alcohol oxidation and also an intermediate in the carbohydrate metabolism. It is used as a standard by which chiral molecules of the D- or L-series are compared. Further oxidation of glyceraldehyde produces carboxylic acids, such as glyceric acid and tartronic acid.

c. **Glyceric acid** (GLA) is used as one of the components for gentle skin peeling.[17] It is also used as a monomer of a polyester packaging material for exothermic and volatile substances.[18] Moreover, in its ester form, it can act as an effective biodegradable fabric softener.[19]

d. **Tartronic acid** can be used as a potentiating agent of adjuvant to increase the blood adsorption of a tetracycline antibiotic.[20] Moreover, it can be used to scavenge dissolved oxygen in alkaline water.[21]

The above mentioned products, especially dihydroxyacetone, can also experience a C-C cleavage producing substances with two atom carbons. Some of these substances are:

a. **Glycolic acid**, the simplest hydroxy acid. It can be used as a raw material for cosmetics and biodegradable polymers.[22]

b. **Oxalic acid**, the simplest dicarboxylic acid. It can be used as reducing agent, as rust removal for metal treatment, as bleaching agent, etc.[23]

2.2 Heterogeneous catalysts for glycerol oxidation reactions

2.2.1 Introduction

An aerobic heterogeneous catalytic alcohol oxidation reaction, which uses air or oxygen as a sole oxidant, was reported for the first time by Edmund Davy in 1820. Moreover, the utilization of heterogeneous catalysts were even mentioned earlier in 1813 by Louis Jacques Thenard.[24] However, investigations in this field have been intensified just 30 years ago due to the increase of environmental concern.

In a selective oxidation reaction, a preferential oxidation of an alcohol functional group in a compound is conducted to convert it into a carbonyl group or a carboxylic group. In earlier works, these reactions were carried out using heterogeneous platinum and palladium catalysts. In these cases, platinum was used either as platinum black or in a dispersed form on a support material.[25]

During selective catalytic oxidation reactions, glycerol is converted to glyceraldehyde if the primary alcohol is oxidized first. On the other hand, dihydroxyacetone (DHA) will be produced if the secondary alcohol group of glycerol is oxidized first. Theoretically, both of these products can be further oxidized resulting in a complex reaction pathway as shown in figure **2.2.1**.

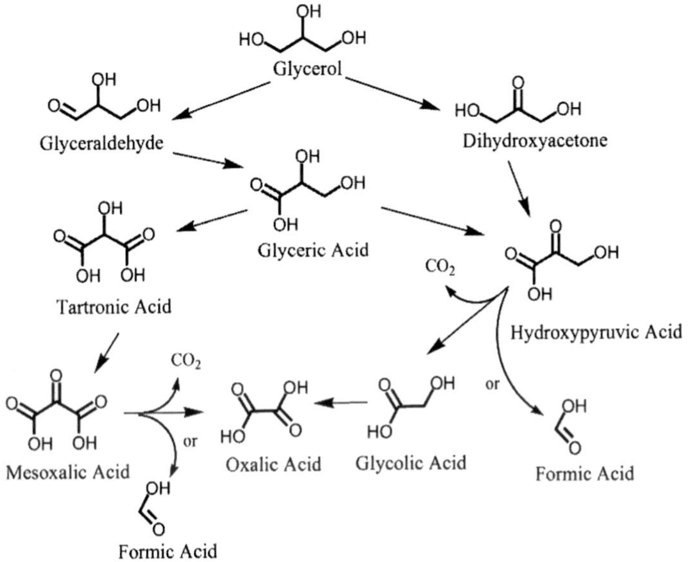

Figure 2.2.1: Possible products of glycerol oxidation reactions.[26-27]

An investigation on the glycerol reaction pathway was started by Prati's group. Based on the theoretical reaction pathways (figure **2.2.1**), following the primary alcohol oxidation pathway mesoxalic acid is an oxidation product from tartronic acid. Furthermore, oxalic acid may originate from mesoxalic acid.[26] However, it was also assumed that oxalic acid is a product from glycolic acids, which may originate from hydroxypyruvic acid.[28] Finally, hydroxypyruvic acid is an oxidation product from dihydroxyacetone following the secondary alcohol oxidation pathway.

2.2.2   Early catalytic glycerol oxidation reactions

The catalytic oxidation of glycerol has been intensively examined since the beginning of the 90's. The earliest work in aerobic aqueous glycerol oxidation was carried out by Kimura's group using palladium and platinum catalysts supported on granular charcoal.[29] While in reactions catalyzed by platinum alone the oxidation towards the primary alcohol was preferred, the addition of a p-electron metal (e.g., bismuth) to the platinum catalyst resulted in an improved oxidation of glycerol's secondary alcohol. For example, when bimetal Pt-Bi used as catalyst, the selectivity towards dihydroxyacetone has been increased from 10 to 80%. In addition, the Kimura's group did not only convert glycerol to its derivates such as glyceric acid and dihydroxyacetone, but also carried out tandem oxidation-amination reactions producing D,L-serine and glycine.[30]

The effect of p-electron metal addition on increasing the selectivity towards glycerol's secondary alcohol oxidation has also been observed by Garcia and Besson.[31] Here, the maximum yield of 30% dihydroxyacetone could be achieved using a platinum-bismuth catalyst instead of an unpromoted platinum catalyst (12% yield). However, the same effect was not observed for palladium-bismuth catalyst, where the addition of bismuth decreased the conversion of glycerol without increasing any selectivity. In addition, the selectivity towards glyceric acid was higher when the reaction was catalyzed by an unpromoted palladium catalyst (selectivity: 77%) compared to unpromoted platinum (61%).

2.2.3   Gold catalysts in alcohol and glycerol oxidation reactions

In spite of its good performance for aerobic glycerol oxidation processes, platinum catalysts are vulnerable from oxygen poisoning especially at high pressure. Therefore, they can only be used at low oxygen partial pressure.[26] On the other hand, gold has shown a better resistance to oxygen poisoning compared to platinum.[32]

Heterogeneous gold catalysts for the selective aerobic oxidation of alcohols were introduced by Prati's group in the late 90's for the oxidation of 1,2-propandiol[32] and ethylene glycol[33]. In their work, the sol-gel method as a new technique in the preparation of heterogeneous catalysts for catalytic alcohol oxidation reaction was utilized for the first time.[33] Afterwards, this method was used also to prepare heterogeneous catalysts for the glycerol oxidation reaction. The advantage of this method was its ability to produce very small metal nanoparticles. For example, gold nanoparticles having particle diameters 4 ± 2 nm[34] or bimetal Au-Pd nanoparticles with 4 nm diameter[35]. The particles produced were usually smaller than those that are prepared by other methods such as by precipitation methods.

The first implementation of heterogeneous gold catalysts for glycerol oxidation was carried out in 2002 by Hutchings et al. and resulted in a yield of 56% towards glyceric acid as the best result.[7] This study also showed that without the presence of NaOH no reaction could take place, whereas, as proposed, the OH⁻ is necessary for the abstraction of the H^+ of a hydroxyl group of glycerol. Furthermore, it was also shown that a high concentration of NaOH led to a high selectivity toward glyceric acid[7] and an increased conversion of glycerol[36]. However, when the concentration of NaOH was too high, the produced glyceric acid was oxidized further to tartronic acid.[36] Finally, the presence of NaOH was also able to reduce the adsorption of poisoning substances on catalysts surfaces.[37]

2.2.4 Oxidation reaction without additional base

Generally, additional base, e.g. NaOH, was added because the atomic O to oxidize glycerol originates from a hydroxide intermediate, while the diatomic O_2 would regenerate the consumed OH⁻. Consequently, alkaline condition would decrease the energy required to carry out the reaction.[38] However, in contrast to others, Liang et al. were able to produce highly dispersed Pt nanoparticles (d < 6nm) supported on activated carbon.[39] This made it possible to carry out the glycerol oxidation reaction without the addition of bases. By treating the activated carbon with H_2O_2, the surface area and the total pore volume of the support could be increased from 1373 to 1492 m^2/g and from 0,88 to 0,93 cm^3/g, respectively. This highly porous support combined with an optimum catalyst preparation method enabled the production of highly dispersed (80% surface coverage) small size platinum catalysts. Results of base free glycerol oxidation experiments are displayed in table **2.2.1**.

Table 2.2.1: Conversion and selectivity of glycerol oxidation in catalytic aerobic base free reactions.

Metals	Conversion	Selectivity [%]			
	[%]	Glyceric acid	Glycolic acid	DHA	Tartronic acid
AuPt/MgO[40]	43	72	3	n.d.	15
Pt/AC[39]	50	47	13	17	n.d.
PtCu/AC[41]	86	70	10	10	n.d.
Au/hydrotalcite[42]	44	n.d.	53	n.d.	n.d.

2.2.5 Metal comparison

The first comparison of platinum, palladium and gold catalysts for glycerol oxidation carried out by Hutching's group showed similar conversions (circa 60%) and selectivities towards glyceric acid (circa 70%) between the palladium and platinum catalyzed reaction, while 100% selectivity towards glyceric acid at 56% conversion was reached using gold catalysts.[43] On the other hand, Bianchi mentioned that at 50% conversion the highest selectivity towards glyceric acid was achieved by using palladium (80% selectivity) and the lower selectivity were achieved by gold (64% selectivity) and platinum (42% selectivity).[44] A comparison of monometal catalysts is displayed in table **2.2.2**. [45]

Table 2.2.2: Comparison between monometallic catalysts supported on activated carbon for the glycerol oxidation.[45]

Metals	Conversion [%]	Selectivity [%]			
		Glyceric acid	*Glycolic acid*	*DHA*	*Tartronic acid*
Pd	95	64	11	16	9
Rh	88	62	15	17	6
Pt	20	60	15	20	4
Au	44	42	20	38	0

Reaction conditions: 60°C, 3 bar O_2, 0.3 M glycerol, NaOH/glycerol=2 mol/mol.

2.2.5.1 Bimetallic systems

The performance of supported mono metal catalysts such as palladium or gold for alcohol oxidation reactions can be improved when they form bimetal catalysts. For example, in the benzylalcohol oxidation to benzaldehyde, the selectivity towards the aldehyde product was increased from 60 to 90%.[46] The performance improvements were also shown in glycerol oxidation reaction. For example, as mentioned by Claus's group, the activity of glycerol oxidation catalyzed by gold increased significantly when it was coupled with platinum metal.[34] Furthermore, it has also been reported that a platinum catalyst modified with gold was more resistant against deactivation.[47]

Bianchi et al. compared bimetal catalysts of Au-Pd and Au-Pt supported on activated carbon for glycerol oxidation. In order to make sure that only the effect of the metal itself was observed, the particle size of the catalysts were made to be similar (2-3 nm) and they were prepared by the same method (sol-gel-method stabilized with PVA). At a conversion of 50% the best catalysts were the Au-Pt catalyst (producing glyceric acid with selectivity of 72%) and the Au-Pd catalyst (with selectivity of 76%) (see table **2.2.3**).[44]

Table 2.2.3: Comparison between monometallic and bimetallic catalysts for the glycerol oxidation at 50°C and 3 bar O_2.[44]

Catalyst	Conversion [%]	Selectivity [%]				TOF [h^{-1}]*
		Glyceric acid	Glycolic acid	Oxalic acid	Tartronic acid	
Pt/C	50	42	31	8	6	532
Pd/C	50	81	3	0	14	1151
	100	21	27	12	39	
Au/C	50	65	12	10	9	1090
	100	45	24	19	10	
Au-Pt/C	50	72	18	1	8	1987
	100	31	35	5	28	
Au-Pd/C	50	77	5	0	18	1775
	100	49	25	2	25	

Glycerol initial concentration was 0.3 M and the molar ratio of NaOH to glycerol was 2.
*Calculation of TOF (turnover frequency) after 0.25 hours of reaction.

The results in table **2.2.3** show that in terms of selectivity the bimetal Au-Pt and Au-Pd have a higher selectivity towards glyceric acid than the mono metal gold and platinum catalysts. However, for the Au-Pt catalyst, prolonged reaction time could cause over oxidation of glyceric acid, which decreased its selectivity and increasing the selectivity of its derivates, e.g. glycolic acid and tartronic acid.[47]

Gold catalysts have a tendency to produce H_2O_2 during the glycerol oxidation process resulting in an increased of C-C cleavage and subsequently an increased selectivity towards C2 products such as glycolic acid and oxalic acid.[28] To improve the selectivity, metallic gold was grown on palladium nanoparticles. For the catalyst where the gold metal covered the whole surface of the palladium there was no significant change in selectivity compared to the mono metal gold catalyst. However, in case the surface of palladium was only partially covered by gold, the production of H_2O_2 was significantly decreased and the selectivity towards glyceric acid was increased (see table **2.2.4**).[48]

Table 2.2.4: Glycerol oxidation and H_2O_2 production rates for Pd, Au and Au-Pd catalysts.[48]

Catalyst	Glycerol Oxidation TOF [h⁻¹] *	H_2O_2 Production TOF [h⁻¹] *	Selectivity to glyceric acid [%]
Pd	3600	-	82
Au	61200	1080	65
Au-Pd (completely covered)	21600	720	64
Au-Pd (partially covered)	7200	108	82

Reaction conditions: 60°C, 10 bar O_2, 0.3 M glycerol and 0.6 M NaOH.
*Calculation of TOF (turnover frequency) after 0.5 hours of reaction.

2.2.5.2. Bimetallic systems: alloys or segregated metals

For the utilization of bimetallic systems it should be also taken into account, whether the bimetals exist as segregated particles or as a bimetallic alloy, since the synergetic effect of bimetals can only be clearly understood when it exists as alloy. Wang et al. proposed a method to synthesis an alloy of Au-Pd catalyst.[49] First, the gold nanoparticles were synthesized and then the palladium nanoparticles were grown on the gold surface by slow reduction using H_2 as a reduction agent. As a result small particles between 3 to 9 nm were formed and each of them exhibited twin boundaries between gold and palladium crystals separated by a 2,29Å lattice spacing, which implies the alloy state. By comparing the performance of the bimetal Au-Pd alloy and the partly segregated bimetal Au-Pd as well as the mono metals Au and Pd, it was shown that the single phase Au-Pd alloy catalyst is the most active and delivers the highest selectivity towards glyceric acid (see Table **2.2.5**).

Table 2.2.5: Selective oxidation of glycerol catalyzed by mono and bimetallic catalysts supported on activated carbon.[49]

Catalyst	Oxidation TOF* [h^{-1}]	Selectivity to glyceric acid [%]
Pd	900	80
Au	1000	68
Au-Pd segregated	4823	74
Au-Pd alloy	6435	77

Reaction conditions: 50°C, 3 bar O_2, 0.3 M glycerol and 1.2 M NaOH.
*Calculation of TOF after 0.25 hours of reaction.

2.2.6 Influence of catalyst particle size

The influence of catalyst particle size on the activity and the selectivity of gold catalyzed glycerol oxidation reactions were observed initially by Carrettin et al. [50] They found that gold particles bigger than 50 nm are not active anymore for this reaction. Therefore, the preparation step to produce small heterogeneous gold nanoparticle catalyst plays a very significant role with respect to the performance of the glycerol oxidation reaction.

Gold nanoparticles with different sizes could be prepared by utilizing different preparation methods, i.e. incipient wetness, impregnation, and sol-gel-methods. By using the sol-gel method stabilized with polyvinyl alcohol, small gold particles of 4-5 nm were produced. They showed a high activity and a medium selectivity (~50%) towards glyceric acid. Bigger particles (> 10 nm) produced by the incipient wetness and impregnation methods showed lower activities. However, they had a higher selectivity towards glyceric acid (> 70%). It was shown that gold particle with a size of approximately 30 nm had the highest selectivity (90%) towards glyceric acid at 90% conversion (see table **2.2.6**).[26]

Table 2.2.6: Glycerol oxidation catalysed with gold particles of different sizes supported on activated carbon.[26]

Preparation method	Particle size [nm]	Glyceric acid selectivity at	
		50 % Conversion	*90 % Conversion*
Impregnation	18	71	70
Incipient wetness	15	80	78
Sol-gel (PVA*)	5	47	35
Sol-gel (sodium citrate)	21	75	75
Sol-gel (sodium citrate) + Calcination	30	90	89

Reaction conditions: 60°C, 3 bar O_2, 0.3 M glycerol and 0.3 M NaOH.
* PVA: polyvinyl alcohol.

The fact that the size of the gold particles correlates with the selectivity for glyceric acid production was also described by other authors.[51] Furthermore, a similar effect was shown for palladium catalysts with particle sizes between 3 to 17 nm. This effect can be observed better at higher temperatures, e.g. 70°C, while at lower temperatures, e.g. 50°C, this effect was not shown.[52]

2.2.7 Effect of preparation methods

The synthesis method to produce metal catalysts plays an important role to determine the activity and selectivity. For example, for bimetal catalysts, it was shown that the simultaneous reduction of the gold and palladium precursors could yield the most active catalysts (e.g., 78% selectivity towards glyceric acid at 50% conversion).[53]

An elaborate comparison of synthesis methods to prepare heterogeneous catalysts for glycerol oxidations was described by Prati.[54] Deposition precipitation, incipient wetness, and sol-gel method are the common methods used to prepare these catalysts. In most cases, investigations were more focused on producing metal particles with a specific narrow range in size between 2-20 nm, then observing their activity and selectivity in glycerol oxidation reactions. Here, reduction of metals plays an important role.

Moreover, the reduction step in producing nanoparticles is not only important to determine the particle size but also the oxidation state of the metals. It was observed by using X-ray photoelectron spectroscopy (XPS) that the Au/TiO_2 catalyst reduced thermally (calcinations to 700°C) still consists of Au^{III} species besides of a majority of Au^0 particles, while the catalysts reduced chemically (e.g., by $NaBH_4$) have all gold as metallic Au^0. Further analysis showed that only Au^0 is the active catalyst for glycerol oxidation.[54]

2.2.8 Effect of reduction methods

The size of nanoparticles depends, among other things, on the nature of the reduction agents. During the formation of transition metal nanoparticles there are two crucial processes which play an important role to determine the particle size, namely nucleation and growth. Generally, if nucleation is faster than growth, particles of smaller sizes will be formed. Therefore, in general, it is necessary to utilize a powerful reductant providing a fast initiation of the nucleation step, which subsequently will lead to the production of small nanoparticles.[55]

For glycerol oxidation reactions, formaldehyde, $NaBH_4$, N_2H_4 and H_2 are the main reduction agents used for the synthesis of nanoparticle catalysts. For the preparation of platinum nanoparticles and gold-platinum bimetal systems, the use of $NaBH_4$ and H_2 as reducing agent resulted in small particles of 2-3 nm, whereas the use of N_2H_4 increased the particle size to 7– 8 nm. For platinum catalyzed reactions, it was shown that higher activity was achieved with small particle sizes ($\leq$ 3 nm). Therefore, the best reduction agent for platinum catalyst is H_2. On the other hand, using H_2 to prepare gold nanoparticle has resulted in a black precipitation. Therefore, $NaBH_4$ was more preferable.[47] A similar observation was described by Esumi et al., that the bimetals Pt-Pd reduced by $NaBH_4$ exhibit particle sizes smaller than the one reduced by N_2H_4.[56]

2.2.9   Effect of support materials

The effect of support material on the selectivity and activity of glycerol oxidation reactions was firstly investigated in detail by Claus' group.[36, 57] A comparison between carbon black, activated carbon, graphite, TiO_2, Al_2O_3, CeO_2 and MgO showed that gold catalysts supported on carbons have a high activity. Furthermore, the carbon black support was more active than activated carbon or graphite. A similar observation was also reported by Prati's group.[54] It was shown that gold metals with a 5 nm size prepared by the same method had higher activity when supported on activated carbon than on TiO_2, with the TOF of 1090 and 178 h^{-1}, respectively.

High activity of catalysts supported on activated carbon was the result of high metal dispersion on this support.[58] This then led to a high surface metal loading.[28] Furthermore, for the carbon based support itself, the activity of the catalysts were mainly influenced by the fraction of micropores (pores with diameter lower than 2 nm). The more micropores were available in a support, the lower was the activity of the catalyst.[34]

Nevertheless, there were several advantages using a metal oxide support. One of them was their contribution to increase the selectivity towards glyceric acid. For example, Au-Pd catalyst supported on TiO_2 showed higher selectivities than the catalyst supported on activated carbon (see table **2.2.7**). This tendency might be influenced by the acidic nature of activated carbon surface. A similar observation was shown also for mono metal gold and palladium catalysts.[35]

Table 2.2.7: Comparison between TiO_2 support and activated carbon for Au-Pd catalysts.[35]

Au-Pd on	Selectivity [%] at 100% conversion				TOF[h^{-1}]*	Particle size [nm]
	Glyceric Acid	Tartronic Acid	Glycolic Acid	Formic Acid		
TiO_2	53	21	11	8	3341	4
activated carbon	44	47	8	-	3999	5,4

Reaction conditions: 60°C, 10 bar O_2, 0.6 M glycerol and 1.2 M NaOH.
*Calculation of TOF after 4 hours of reaction.

Another disadvantage of metal oxides as support material was metal leaching, which consequently decreased the activity of catalyst. Furthermore, it was shown that the performance of the metal oxide catalysts were not significantly influenced by different type of preparation methods, in opposite to carbon supported catalysts.[57]

2.2.10 Glyceric acid as the main product of glycerol oxidation reactions

As a brief review from major reports in this field, the conversion and the selectivity of glycerol oxidation to produce glyceric acid are summarized in table **2.2.8**.

While most studies tried to increase the selectivity of glycerol oxidation reaction towards glyceric acid, Sankar et al. tried to increase the selectivity towards glycolic acid. It has been reported that the highest selectivity (56%) could be reached when a 1% Au/graphite catalyst was used in combination with H_2O_2 as oxidation agent.[28]

Table 2.2.8: The best conversions and selectivities towards glyceric acid catalyzed by Pd, Pt, Au and bimetallic Au-Pd.

Catalyst	Conversion [%]	Selectivity [%]	References
Pd/AC	90	77	[31]
Au/AC	56	100	[43]
Pt/AC	50	47	[39]
Au-Pd/TiO$_2$	100	52	[35]
Au nano colloids	90	76	[59]
Au/ TiO$_2$	33	64	[38]

AC: Activated Carbon.

2.2.11 Glyceric acid as a starting material for further oxidation reactions

Further oxidations of the first generation of glycerol oxidation products were investigated by Gallezot et al.[5] Here, due to its high stability glyceric acid was preferably chosen over dihydroxyacetone as the starting material to produce tartronic acid and hydroxypyruvic acid. By using platinum-bismuth catalyst similar to Kimura's, it was found that tartronic acid can be produced from glyceric acid with a high yield (82%) in basic media (pH = 11). Moreover, at acidic conditions (pH = 3) a high yield of hydroxypyruvic acid (63%) can be reached.[60] This work was continued by van Bekkum investigating the optimum process to produce hydroxypyruvic acid using glyceric acid as starting material.[61] Here, it was found, that at a pH slightly higher than neutral (pH=8) a higher yield (80%) towards hydroxypyruvic acid could be achieved. In case glycerol was used as a starting material, the yield towards hydroxypyruvic acid reached only 20%. A summary of these studies is given in table **2.2.9**.

Table 2.2.9: Further oxidation of glyceric acid.

Catalyst	Product	Conversion [%]	Selectivity [%]	pH	Reference
Pt-Bi/AC	Tartronic acid	92	90	10-11	[60]
Pt-Bi/AC	Hydroxypyruvic acid	67	95	2	[60]
Pt-Bi/AC	Oxalic acid	90	45	decreased from 6 to 3	[61]
Pt-Bi/AC	Hydroxypyruvic acid	100	80	8	[61]
Pd/AC	Tartronic acid	Yield: 70%		11	[5]
Pt-Bi/AC	Hydroxypyruvic acid	77	74	2	[5]

AC: Activated Carbon.

2.3 Immobilized catalysts and their application in catalytic alcohol oxidations

2.3.1 Introduction

Due to their simple and very small structure, molecular or homogeneous transition metal catalysts (TMC), either as an ion or as a complex, can be synthesized and reproduced easily with high accuracy. The electronic and steric properties of these molecules can also be modified easily to improve the performance of a reaction, e.g. by adding a different type of ligand, the electron acceptor in transition metal complexes. For instance, in the cyclic oligomerisation of butadiene catalyzed by nickel, when the ligand tricyclohexylphosphine (PCy₃) was used, 4-vinyl-1-cyclohexene was formed as the main product. However, when triphenylphosphite (P(OPh)₃) was used as the ligand, then 1,5-cyclooctadiene will be produced predominantly.[62]

Unfortunately, in many cases, TMC exhibit a difficult separation from the reaction mixture, which inhibits their utilization at an industrial scale due to the necessity to recover the costly and highly specialized catalysts. Therefore, a wider spread of application may be found if TMC can be immobilized onto easier separable supports, in other words, heterogenization of TMC.

Similar to other heterogeneous catalysts, typical supports to immobilize homogeneous catalysts include metal oxides, such as silica and alumina, zeolites, and carbon based materials such as activated carbon, organic polymers. On these supports, the TMC, either as an ion or a complex, are immobilized via mechanisms such as physisorption, covalent bonds, ion exchange or physical entrapment.[63] Unlike metal nanoparticles, which exist in the oxidation state zero, most of TMC exist with different oxidation states. Moreover, since most metals in their highest oxidation state have the lowest affinity for the support materials, it is difficult to keep the TMC stable on their immobilizing matrix. In addition, three challenges faced in immobilization of TMC were:

• to understand the interaction between catalysts and their matrix,

• to bridge engineering and molecular science, and

• treating immobilization steps as an integral part in catalyst design from beginning.[64]

2.3.2   Methods to immobilize TMC

Despite numerous publications and patents showing the potential of utilizing immobilized TMC in laboratory scale,[65] only very few industrial examples can be listed, e.g. carbonylation of methanol to produce acetic acid catalyzed by $[RhI_2(CO)_2]^-$ electrostatically bound to an ion exchange resin (ACETICA®Process form Chiyoda Corporation)[66]. Moreover, the condition was worse for selective oxidation reactions, since only very few examples, even at laboratory scale, were available. On the contrary, epoxidation reactions by using a metal-salen complex immobilized via different methods were reported in several studies. Considering a relative similarity between epoxidation and oxidation reaction, in this dissertation, different immobilization methods of a well-known catalyst for epoxidation reactions, i.e. metal-salen complexes such as the Jacobsen's catalyst (see figure **2.3.1**) are mentioned to describe the heterogenization of TMC.

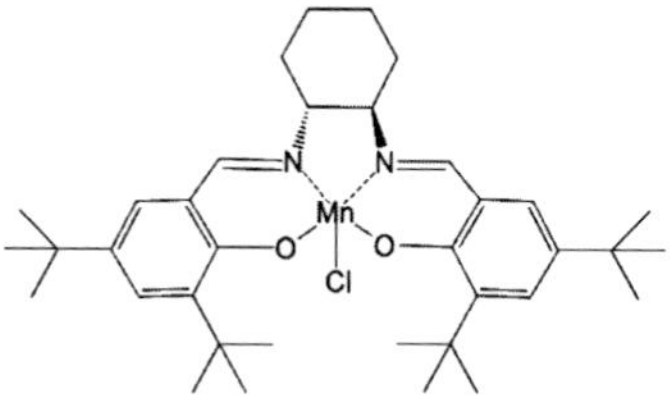

Figure 2.3.1: Jacobsen's catalyst.[67]

Methods that represent the majority of TMC immobilizations can be classified into three groups as follows:[68]

I.   Entrapment

In this method, the TMC were immobilized by encapsulation either in well defined porous materials (e.g. "ship-in-a bottle" approach) or in a polymer matrix.

II.   Grafting and anchoring

In grafting, the metals were linked directly to the surface of the supports, while in anchoring the bonds between metal and support material are facilitated by a linker. The supports include carbon based materials (also polymers) or metal oxides.[69]

III.   Adsorption

This method covers all non-covalent bonding on the surface of porous materials. Ionic binding belongs to this group. In the following few examples of support materials for immobilized metal-salen complexes in epoxidation or oxidation reactions are given:

a.   Montmorillonite

Montmorillonite consists of aluminum-silica layers that form tetrahedral and octahedral nets.[70] Here, metal-salen complexes can be inserted into the space between the silicate layers which offers the diffusion only in two-dimensional space. Therefore, this approach will increase the contact frequency between the reactants and the metal catalysts. For example, in the oxidation of p-cresol catalyzed by cobalt-salen complexes, the immobilized catalyst in montmorillonite displayed an activity about five times higher than the activity of the free catalysts.[71]

b.   Activated carbon

Similar to other heterogeneous catalysts, activated carbon is frequently used as support material due to its high pore volume. The immobilization of metal-salen complexes, e.g. a copper-salen, onto activated carbon might take place via two different mechanisms. One is by a direct bonding between the metal center and the superficial oxygen atoms of the carbon support. Two is by a covalent bond between a modified salen ligand and the oxygen functional groups on carbon outer layers. In the latter case, the bonding of the metal complex was more stable.[72]

c.   Polymer resins

Jacobsen's catalyst had been anchored on a polymeric support and was used as catalyst for olefin epoxidation reactions. For example, a high activity (a yield of 49%) and a high enantioselectivity (91% ee) were displayed in an asymmetric epoxidation of 1-phenylcyclohexene when the catalyst was immobilized on methacrylate-based resin. This result reached almost the performance of the non immobilized catalyst, i.e. 72% yield and 92% ee.[73]

d.   Silica supports

Silica supports together with polymers were the most mentioned supports in the immobilisation of TMC (figure **2.3.2**). For example, a palladium Schiff base catalyst supported on silica gel was used in the oxidation of benzyl alcohol to benzaldehyde with 95% yield.[74]

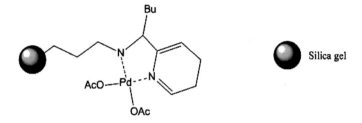

Figure 2.3.2: Palladium-Schiff base complex anchored on silica gel.

e.   Zeolite MCM-22

The disadvantage of anchoring methods as displayed by polymer or silica supports was the necessity to modify the ligands, thus, changing the properties of the catalyst complexes. Even though, by using physical adsorption methods no ligand modification was needed, but the bond between the catalyst-complex and the support material was usually not stable. On the other hand, when the complex was entrapped inside a pore of supports such as zeolites or mesopore materials, then a stable immobilization could be achieved without any ligand modification. Similar to the Montmorillonite entrapped catalyst, the enantioselectivity of the immobilized catalysts could be higher than the free catalysts due to the steric constraint imposed by the framework. For instance, in the epoxidation of α-methylstyrene catalyzed by Jacobson's catalyst, the enantioselectivity of the reaction that was catalyzed by the catalyst entrapped in MCM-22 (see figure **2.3.3**) was significantly higher than the free catalyst (91% as against 51% ee).[75]

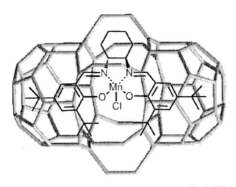

Figure 2.3.3: Schematic Jacobsen's catalyst entrapped in zeolite MCM-22.

2.4   Metal nanoparticles

2.4.1   Introduction

The utilizations of catalyst nanoparticles have been intensively developed for the last 10 years. The terminology of nanotechnology itself was introduced by Norio Taniguchi in 1974, referring a technology to produce material in ultra fine dimension (in the order of 1 nm) with extra high accuracy. Further, NASA described it as the creation of materials through control of matter on the nanometer length scale (1–100 nm).[76] Due to their very small size, catalyst transition metal nanoparticles exhibit very large surface area to volume ratio, accordingly, having a very high surface reactivity especially compared to bulky heterogeneous catalysts. With only very few examples, the applications of unsupported nanoparticle catalysts in glycerol oxidation reaction were also reported.[59, 77]

A cluster of transition metal nanoparticles consists of tens to hundreds of atoms. Metal clusters themselves have a tendency to build bigger particles; therefore, they have to be protected to avoid agglomeration. In the liquid phase preparation methods, the most common stabilizing agents were:

a. Polymers

Through the steric bulk of their framework, polymers could stabilize metal nanoparticles. In addition, they could also weakly bind nanoparticle surfaces similar to ligands.[78] One of the most utilized polymers was poly(N-vinyl-2-pyrrolidone) (PVP) because of its dual stabilizing properties, i.e. steric and ligand like. For example, it was used as a chelating agent to stabilize Cu nanoparticles in a polyol process, a nanoparticle synthesis method which uses alcohols as stabilizing or reducing agents.[79] Other examples of stabilizing polymers were polyacrylic acid, chitosan, oligosaccharides, etc. [65]

One application of this approach was, for example, by using gold-platinum catalysts stabilized in a cross-linked polystyrene derivative for aerobic alcohol oxidations. Thus, the oxidation reactions could be carried out at very moderate conditions, i.e. at room temperature and without additional base. Furthermore, all reactions that used either primary aromatic or allylic alcohols as starting materials led to the excellent production of aldehydes without the production of carboxylic acids.[80]

b. Dendrimers

Dendrimers are similar to polymers but more perfectly defined. They form shapes like cauliflower and become globular after several generations. Therefore, they can entrap metal nanoparticles in their interiors. Their branched structures could also serve as filters for molecules to access the catalysts inside.[78] A well-known dendrimer in this field is poly(amidoamine) or PAMAM. Since the terminal groups of this dendrimer could be modified easily, it properties such as solubility and molecular selectivity can be simply varied. It was utilized, for example, in the selective hydrogenation of allylic alcohols,[81] for the conversion of resazurin to resorufin,[82] in the reduction of p-nitrophenol,[83] etc.

Unfortunately, only few studies reported the utilization of metal nanoparticles in catalytic alcohol oxidation reactions. The dendrimer poly-2-(methylthio)ethyl methacrylate–N,N-dimethylacrylamide–N,N'-methylenebisacrylamide (MTEMA-DMAA) stabilized gold-palladium nanoparticle catalyst was one example of this system. This catalyst was used in the oxidation of n-butanol and could deliver a relatively higher conversion than a Au-Pd catalyst on activated carbon. Moreover, a high selectivity towards n-butanal could be achieved by using this dendrimer stabilized catalyst, whereas a high selectivity towards n-butanoate could be achieved by using the catalyst supported on carbon.[84]

c.   Ligands

The most precise synthesis of metal nanoparticles perhaps can be achieved by utilizing ligands as stabilizing agents. A first example of ligands as nanoparticle stabilizer was the utilization of citric acid to produce gold nanoparticles. Furthermore, organic ligands like phosphine groups were often used to stabilize metal nanoparticles.[85] A typical example was the gold cluster nanoparticle $Au_{55}(PPh_3)_{12}Cl_6$.[86] In addition, another advantage by using ligands as stabilizing agents was the possibility to carry out enantiomeric reactions such as the enantioselective allylic alkylation of *rac*-3-acetoxy-1,3-diphenyl-1-propene with dimethyl malonate (see figure **2.4.1**). This reaction was catalyzed by palladium nanoparticles stabilized by chiral xylofuranoside diphosphite ligands.[87]

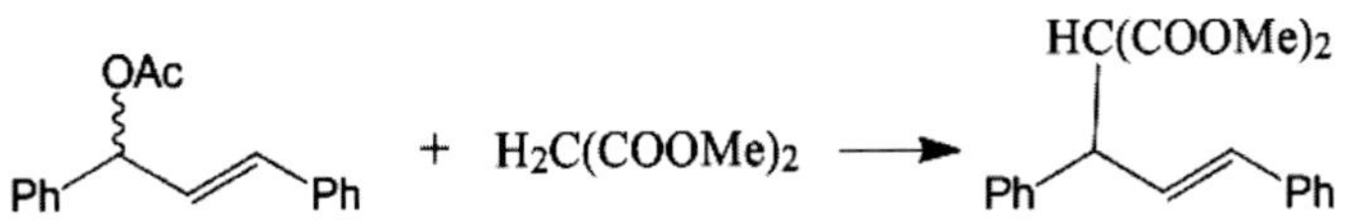

Figure 2.4.1: Asymmetric allylic alkylation of *rac*-3-acetoxy-1,3-diphenyl-1-propene with dimethyl malonate.[87]

Beside the three most common stabilization methods which were mentioned before, surfactants (e.g. oleic acid) can also be used in the formation of metal nanoparticles. This type of method is well used for the synthesis of magnetic nanoparticles. Furthermore, another stabilizing method is by fixing metal nanoparticles onto a porous heterogeneous support such as carbon based materials and metal oxides. Thus, these heterogenized nanoparticle catalysts can be recycled like other bulky heterogeneous catalysts.[88]

2.4.2   Synthesis of metal nanoparticles

Methods to produce metal nanoparticles catalysts can be categorized as physical and chemical methods. Some examples of physical methods were sputtering, evaporation (thermal and electron beam), pulse laser deposition (PLD), and ion implantation. Further, some examples of chemical methods were sol–gel, co-precipitation, impregnation, and chemical vapor synthesis (CVS).[89] In this report, only few examples of the chemical methods which were commonly used to prepare heterogeneous catalysts for oxidation reactions will be discussed.

2.4.2.1   Sol-gel method

The sol-gel method was started by formation of a stable colloidal solution. This was achieved by dissolving precursor transition metals (e.g. $NaAuCl_4 \cdot 2H_2O$, Na_2PdCl_4, or K_2PdCl_4[44]) in a solvent in the presence of a stabilizing agent, such as polyvinyl alcohol (PVA), poly(vinylpyrrolidone) (PVP) or tetraalkylammonium surfactant, which provides a stabilization by means of electrostatic or steric effects. The precursor was then reduced by a reducing agent, e.g. N_2H_4, or $NaBH_4$), to produce a colloidal solution or a "sol". This step was followed by the condensation of the colloidal particles to produce polymeric chains in the solution. Thus, it resulted the formation of a "gel" (e.g. liogel or hydrogel) (see figure **2.4.2**).[90] However, in the preparation of heterogeneous catalysts for glycerol oxidation reactions, usually this method was carried out only until the formation of sols because they are adsorbed further on a supporting material such as activated carbon, metal oxides, etc.

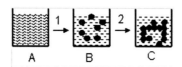

Figure 2.4.2: Flow scheme of sol-gel preparation method.   **A.** Precursor solution; **B.** Colloidal solution; **C.** Gel in solution.[90]

Sol-gel methods were very favorable for the production of multi metallic nanoparticles especially magnetic particles due to their simple procedures.[91] Furthermore, these methods enabled the reduction of the precursor metals to be carried out either simultaneously (producing a core/shell structure or an alloy structure) or successively (producing only a core/shell structure).[92]

One application of the sol-gel method was in polyol processes. Here, alcohol groups played a significant role in the synthesis of nanoparticle. They covered usually the utilization of diols such as 1,2-hexadecandiol[93-95], 1,2-propanediol[91, 96], ethylene glycol,[91, 97-101] etc. as reducing agents. Usually these processes were combined with an addition of protecting agents such as oleic acid, oleylamine[95] and poly(vinyl pyrrolidone)[79, 101-102] to avoid agglomeration of the produced nanoparticles. Furthermore, ascorbic acid could also be added as an antioxidant.[79] Recently glycerol's ability to reduce metal precursors has been shown in the production of platinum nanoparticles assisted by microwave.[100]

2.4.2.2 Deposition-precipitation

In a deposition-precipitation technique, a catalyst precursor (in a solution) can be deposited onto a surface of supporting material by precipitation. Here, during the precipitation, the agglomeration of precursor metals to a bulk metal in the solution should be prevented to avoid producing inactive catalysts. Precipitation of the precursor metals can be done by, for example, changing the pH of the solution, reducing the precursor metals, ligands or complexing agent removal (e.g., volatilisation of ammonia from amine complexes), etc. Impregnation methods and co-precipitation methods belong to this group.[103]

a.   Impregnation

In the impregnation method, a solution of precursor metal was added to a support material dispersed in a solvent. Then, the adsorbed precursor was reduced to produce metal nanoparticles. One example was by adding a chloroauric acid ($HAuCl_4$) solution to an activated carbon suspension followed by a reduction with formaldehyde.[104] A variant of this method was incipient wetness.

b.   Precipitation and coprecipitation

During precipitation, metal nanoparticles were separated from the homogeneous solution after a super saturation condition had been reached. Usually this method was used to prepare support materials for catalysts. Coprecipitation was defined as the simultaneous precipitation of several components. It was used, for example, in the production of an alumina-supported nickel catalyst Ni/Al_2O_3. Here, a pure hydrotalcite-like structure was formed without any formation of nickel hydroxides (no nickel segregation). This could be achieved by very effective mixing of the catalyst constituents on an atomic scale.[90]

2.5 Magnetic particles and their application in processing industry

2.5.1 Separation of magnetic particles

Through magnetic separations, particles were separated based on their different attractiveness in a magnetic field. The equipment used in the separation of magnetic material from the liquid phase included:[105]

a.   wet magnetic drum separators

b.   magnetic filters

c.   wet high-intensity magnetic separators (WHIMS)

d.   high-gradient magnetic separators (HGMS)

When the particles to be separated had only low magnetic properties (paramagnetic particles), then it was necessary to create a high magnetic field gradient (figure **2.5.1.a**). Hence, one could limit the separation volume, e.g. by using smaller containers. Furthermore, by introducing some small ferromagnetic elements within the separation volume, one could create a high local gradient of magnetic field. Consequently, that allowed the creation of very intense magnetic forces with short-range action (figure **2.5.1.b**), which was the characteristic of HGMS. Therefore, this system was particularly important for capturing very small particles with low magnetic properties. For example, it was applied to purify kaolin clays, for the desulphurisation of coal, to treat water polluted with heavy metal ions, and to purify industrial gases.[106]

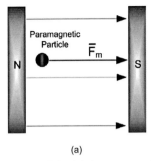
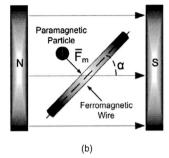

(a)   (b)

Figure 2.5.1: **(a)** A paramagnetic particle in a separation volume under magnetic force Fm. **(b)** The introduction of a ferromagnetic wire creates a higher local magnetic field gradient.

2.5.2   Magnetic nanoparticles and their application in catalysis

Due to their very small size, the applications of magnetic nanoparticles are particularly attractive, e.g. as catalyst supports or in drug delivery. Moreover, they also have a much higher magnetization (per atom) than bulky magnetic materials. Furthermore, magnetic nanoparticles have a dual paramagnetic-superparamagnetic behavior.[107]

Most of magnetic materials were either metals or metal oxides. Therefore, their syntheses did not differ so much from other nanoparticle preparation methods. Two most important steps in these syntheses were controlling the kinetics of nucleation and the growth of the nanocrystals. After the crystal grow, it was important to restrict their agglomeration. This could be achieved by introducing stabilizing agents such as coordinating polymers, capping ligands or dendrimers.[108]

Besides their big potential, it was quite surprising that the utilization of magnetic nanoparticles in catalytic processes is still very limited. Few examples of magnetic nanoparticles in catalytic reactions are presented in table **2.5.1**.

Table 2.5.1: Utilization of magnetic nanoparticle supports in catalytic reactions.

Type of reaction	Catalyst	Remarks
Oxidation of cyclic amines [109]	Au on CeO$_2$/FeO$_X$	The best performance was shown when the educt was dibenzylimine (appox 90% yield), but the conversion was 8% decreased after one time recycling.
Oxidation of cyclic alcohols and olefins [110]	Pd on Fe$_3$O$_4$	The oxidation of benzyl alcohol showed high TON (> 700) and high selectivity (99%). Catalysts were reused for five times and showed insignificant changes.
Oxidation of cyclohexane and toluene [111]	CoFe$_2$O$_4$	After five times used, the conversion (approx. 12%) and selectivity (approx. 90%) were relatively stable.

2.6 Reactor types for heterogeneous catalytic reactions

Common reactor types used for heterogeneous catalytic reactions are fixed-bed, fluidized bed and slurry reactors (see figure **2.6.1**). In a fixed-bed reactor, gas or liquid reactants are passed through a bed of catalysts, thus this system is considered as the simplest type of reactor for heterogeneous systems. However, high pressure drops usually become the biggest disadvantage of fixed bed reactors. Therefore, considering this aspect, the utilization of very small catalysts is impractical. Unfortunately, this leads to lowering the catalyst effectiveness.[112]

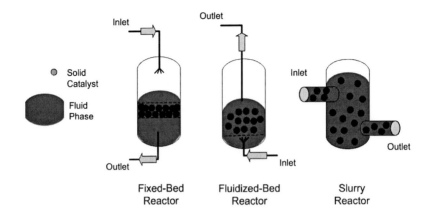

Figure 2.6.1: Three different reactors for heterogeneous reactions.

In a fluidized-bed reactor, an initially fixed-bed was made to a "fluidized" state by an upward stream of gas or liquid. The advantages of a fluidized-bed compared to a fixed-bed were better gas-solid contacts, intensive mixing and better heat and mass transfers. However, the equipments of fluidized-bed reactors were usually more complicated and expensive. Further, the selectivity might be low due to back mixing and bypass.[113]

With respect to heat distribution, the performance of fluidized-beds displays a superiority compared to the fixed-bed reactors. This statement was confirmed by a comparison study such as conducted by Gallucci during a methane reforming process. Consequently, the reaction rate was higher and the heat utilization was more efficient for the fluidized-bed.[114]

A fixed-bed reactor might be used in combination with a fluidized-bed reactor to improve its performance because the dual reactor system can synergize the advantages of both reactors. For example, a simulation of a Fischer-Tropsch-Synthesis of combined fixed- and fluidized-bed reactors showed a 46% rise of yield towards C5+ products compared to a single fixed-bed reactor. This improvement was achieved because the partially converted mixture from the fixed-bed had been further converted in the fluidized-bed.[115]

Another type of non fixed-bed heterogeneous catalytic reactor system is the slurry reactor. Here, the catalysts are finely dispersed in a liquid medium either by mechanical or gas-induced agitation. Therefore, the catalyst could be evacuated together with the reaction medium from the reactor, thus, this might be advantageous with regard to catalyst recycling.

Some industrial application of slurry reactors included hydrogenation of unsaturated oils, Fischer-Tropsch hydrocarbon synthesis, olefins oxidation, etc.[116] Some advantages of a slurry reactor compared to a fixed-bed reactor are:[117]

a. Possibility to utilize smaller catalyst particles
b. Higher mass transfer coefficient
c. Higher heat transfer coefficient
d. No partial wetting of catalyst can take place
e. Better temperature control[112]

Moreover, the utilization of a slurry reactor is more suitable if the catalysts should be recycled frequently.

On the other hand, compared with fixed-bed reactors, slurry reactors are at a disadvantage especially because of the following aspects:[118]

a. Additional investment costs for catalyst separators, e.g. hydrocyclones or filters.
b. Lower catalyst loading capacity due to the maximum catalysts concentration to avoid the suspension from settling
c. Backmixing which reduces the selectivity of the reaction
d. More complex design[112]

In case of the Fischer-Tropsch reaction, a less effective heat transfer in a fixed-bed reactor might result in hot spots. This would eventually decrease the chain growth and increase the selectivity of methane formation. On the other hand, the slurry reactor could deliver a higher selectivity towards C5+ products not only because of better heat transfer, but also due to higher solubility and contact time of the reactants (H_2 and CO).[119]

3   RESULTS AND DISCUSSION

3.1   Heterogeneous catalysts for glycerol oxidation reaction

The evaluation of heterogeneous catalysts for glycerol oxidation reactions (figure **A5.1** in appendix **A5**) were conducted in an pressurized stirred tank reactor (figure **A3.2** in appendix **A3**) and the reaction conditions are mentioned in section **5.3.1**. To begin with, the performances of some well-known commercial catalysts for glycerol oxidation were re-examined (section **3.1.1**). This was followed by more detailed observations on the effect of reaction conditions, such as pressure (section **3.1.2**), temperature (section **3.1.3**), as well as initial glycerol and NaOH concentrations (sections **3.1.4** and **3.1.5**).

3.1.1   Catalysts comparison

In the catalysts screening test, the performances of the well-known commercial catalysts for alcohol oxidation reactions, i.e. palladium or platinum metals on graphite or Al_2O_3 were compared. To guarantee that only the effect of the metal types was observed, the concentration of metals in each catalyst was kept the same, i.e. 1 wt%. The important parameters to be observed were the reaction conversion and selectivity (see appendix **A1** for more detail). Moreover, since the catalysts were introduced in the form of pellets, their mechanical stability was also considered an important parameter.

The conversion of glycerol to C2 products (substances with two carbon atoms) such as oxalic or glycolic acids might lead to the production C1 products, for example formic acid or CO_2 (see figure **2.2.3**). Therefore, from the atom economy point of view producing C3 products such as glyceric or tartronic acids was more desirable. The results of the catalyst screening are depicted in table **3.1.1**.

The conversion behaviour of palladium and platinum catalysts supported on the same material, as presented in table **3.1.1**, indicated that the type of metal influences the performance significantly. Moreover, the product selectivity also demonstrated a disparity between both metals. Figure **3.1.1** shows the yields of glyceric acid over time from the reactions catalyzed by palladium and platinum on the same support material (Al_2O_3).

Table 3.1.1: Initial catalysts screening for batch glycerol oxidation reaction.

Catalyst	Pressure [bar]	Temperature [°C]	Conversion [%]	Selectivity (%)		
				Glyceric Acid	Oxalic Acid	Tartronic Acid
Pd/Al$_2$O$_3$	10	80	44	78	5	16
Pt/Al$_2$O$_3$	10	80	35	85	4	19
Pd/Graphite	10	80	44	79	5	17

Initial reaction conditions: 0.6 M of glycerol and 0.8 M of NaOH.

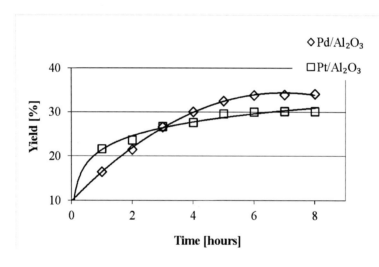

Figure 3.1.1: Yield of glyceric acid with different metal catalysts. Reaction conditions: 80°C, 10 bar air, 0.6 M glycerol, 0.8 M NaOH.

At the beginning, the yield of the palladium catalyzed reaction was slightly lower than with platinum. However, then the yield of the former was increasing more rapidly and after three hours it reached the similar level as the latter. Moreover, after four hours the palladium's yield was even higher than the platinum's, which had already reached its maximum value and did not increase any more. Finally, after six hours the yield of palladium catalyzed process reached the maximum value.

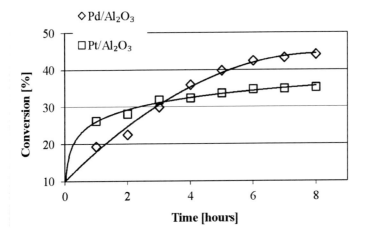

Figure 3.1.2: Glycerol's conversion catalyzed by Pd and Pt metals on Al_2O_3. Reaction conditions: 80°C, 10 bar air, 0.6 M glycerol, 0.8 M NaOH.

Observing the conversion profile of both catalysts enabled a further explanation of their behaviour. As presented in figure **3.1.2**, even though showing higher values in the beginning, the conversion of the platinum catalyzed process did not increase significantly anymore after four hours of reaction and it reached a lower end value than with palladium. Thus, it could be deduced that the platinum catalyst is not as stable as the palladium catalyst in a longer time despite its higher initial activity. Therefore, it was assumed that the lower yield towards glyceric acid of the platinum catalyzed process is mainly due to catalyst deactivation.

Contrary to the type of metals, the type of support materials did not influence the performance of the reaction (see figure **3.1.3**). However, with respect to another criterion, i.e. mechanical stability, the aluminium oxide displayed a better performance than the graphite support. While some of the graphite pellets were destroyed after eight hours of mechanical stirring, the form of the Al_2O_3 pellets were still intact even after 48 hours of reaction. Due to their superior performance as described before, the Pd/Al$_2$O$_3$ was chosen as the best catalyst for further investigation. Figure **3.1.3** displays the similarity between the yield of the palladium catalyzed reactions supported on Al$_2$O$_3$ and graphite.

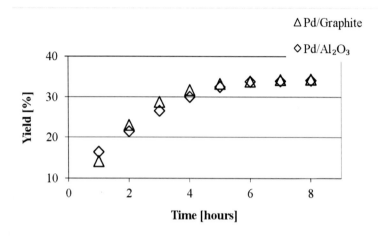

Figure 3.1.3: Yield of glyceric acid with different support materials. Reaction conditions: 80°C, 10 bar air, 0.6 M glycerol, 0.8 M NaOH.

3.1.2 Influence of reaction pressures

Initially, trials were made to develop a process that can run at one atmospheric pressure. After it had been found that the atmospheric batch-wise system cannot deliver any significant conversion, oxygen or synthetic air was introduced continuously to the reactor. For this purpose, two different methods to supply oxygen for the reaction were developed and compared, i.e. via a trickle-bed and a fluidized bed system that can be seen in figures **5.3.2** and **5.3.3**.

Unfortunately, both systems could also not deliver a significant result. Even with good aeration system, the amount of O$_2$ involved in the reaction was limited by the low solubility of O$_2$ in the reaction medium at atmospheric pressure. Therefore, it was concluded that applying pressure higher than one bar air is necessary for the aerobic catalytic glycerol oxidation.

Further experiments were carried out to observe the effect of pressure on the glycerol oxidation process. The results of these experiments at the pressures between 6 to 12 bar are shown in table **3.1.2**.

Table 3.1.2: The effect of air pressure on the performance of batch glycerol oxidation reactions. Reaction conditions: 80°C, 8 hours of reaction, 0.6 M of glycerol, 0.8 M of NaOH, catalyzed by Pd/Al$_2$O$_3$.

Pressure [bar]	Conversion [%]	Selectivity (%)		
		Glyceric Acid	Oxalic Acid	Tartronic Acid
12	36	76	15	6
10	44	79	3	17
8	32	85	3	11
6	20	96	-	3

Since oxygen solubility in an aqueous glycerol solution increases with the rise of pressure at a constant temperature, higher reaction conversion could be reached by increasing the pressure in the reactor. This was confirmed by table **3.1.2**. However, increasing the pressure higher than 10 bar air did not result in further conversions. It was assumed that the catalyst activity is reduced due to oxygen poisoning. Consequently, 10 bar air was chosen as the standard pressure in further experiments.

3.1.3   Influence of reaction temperatures

To investigate the effect of temperature on the performance of the catalytic reaction, the glycerol oxidation catalyzed by Pd/Al$_2$O$_3$ was carried out at temperatures ranging from 30 to 140°C. However, because it was found in the beginning of the experiment that at temperatures between 30 and 40°C no significant glycerol conversion can be observed, this low temperature range was excluded from further experiments.

At temperatures of 60°C and higher, the oxidation reactions displayed significant conversion. Furthermore, a higher reaction rate was observed when the temperature was increased from 60 to 80°C. While at 60°C the conversion rose very slowly, at 80°C the reaction required only 5 hours to reach its near-maximum conversion (figure **3.1.4**). Due to its very slow reaction rate, the temperature of 60°C was excluded from further tests and the performance of reactions at 80 and 100°C was discussed below.

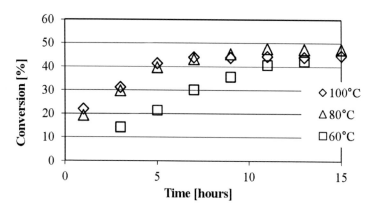

Figure 3.1.4: Glycerol conversion at different temperatures. Reaction conditions: 10 bar air, 0.6 M of glycerol, 0.8 M of NaOH, catalyzed by Pd/Al$_2$O$_3$.

Raising reaction temperature from 80 to 100°C, contrarily, did not lead to a significant increase of glycerol conversion (figure **3.1.4**). Although a slight higher conversion of the reaction at 80°C could be seen after 11 hours, both reactions had generally similar conversion values. A thoughtful observation was made to examine the effect of temperature on production of glyceric acid.

The effect of raising temperatures on the production of glyceric acid is highlighted in table **3.1.3**. A decline in selectivity as a result of the temperature increase from 80 to 100°C was shown. This decline was followed by increase of selectivity toward tartronic acid, an oxidized product of glyceric acid.

Increasing the reaction performance led to the further oxidation of glyceric acid, thus lowering its selectivity. Although considered having a higher economic value than glyceric acid, however, tartronic acid has also a chelating property. Unfortunately, this chelating substance might be adsorbed strongly on the catalyst surfaces, thus decreasing the activity.

Table 3.1.3: The effect of temperature on the performance of batchwise glycerol oxidation reactions. Reaction conditions: 10 bar, 0.6 M of glycerol, 0.8 M of NaOH, catalyzed by Pd/Al_2O_3.

Temperature [°C]	Conversion [%]	Selectivity (%)	
		Glyceric Acid	*Tartronic Acid*
100	44	60	38
80	45	79	17
60	36	86	13

Moreover, further increasing reaction temperature to 120°C and higher caused a significant catalyst deactivation, probably due to chelating substance poisonings. Therefore, 80°C was chosen as the reaction temperature in further experiments.

3.1.4 Influence of glycerol initial concentration

In order to minimize the reactor volume, it is necessary to feed the reactor with glycerol solution at the highest possible concentration. Since in most reports on this field, the solution was introduced only at low concentration, the feasibility of the aerobic catalytic process with high glycerol concentration should be evaluated. Therefore, the effect of glycerol initial concentration, in a range between 0.3 to 1.6 M, towards the reaction performance was investigated.

A better result of glycerol oxidation reactions is shown with low concentration of glycerol in aqueous substrate solutions. As it can be seen from figure **3.1.5** the conversion and the yield towards glyceric acid were higher when glycerol was introduced with concentrations of 0.3 and 0.6 M, compared to concentrations of 1.0 and 1.6 M. This fact might be caused by the catalyst deactivation due to the excessive glycerol supply. The excess of glycerol might fully cover the catalyst active sites, thus making them inactive. Therefore, for further experiments, the concentration of 0.6 M was chosen as the initial glycerol concentration.

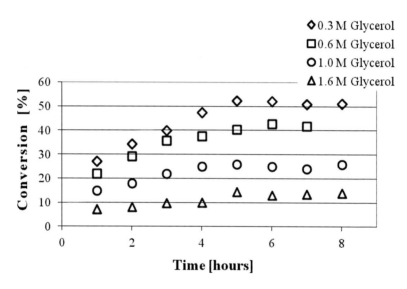

Figure 3.1.5: The effect of glycerol initial concentration towards conversion. Reaction conditions: 10 bar air, 80°C, 0.8 M of NaOH, catalyzed by Pd/Al_2O_3.

3.1.5   Influence of alkaline concentration

As already mentioned in section **2.2.3**, in general, aerobic catalytic glycerol oxidation should be carried out in an alkaline condition (mostly by adding NaOH). Nevertheless, very limited cases showed that the reaction can run also in a base free condition, for example by using very small heterogeneous platinum catalyst ($\leq$ 6nm) mentioned in section **2.2.4**. Here, the effect of alkaline condition (addition of NaOH) towards the performance of the reaction was reevaluated.

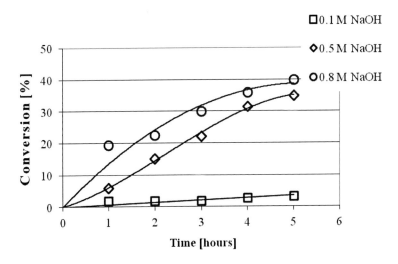

Figure 3.1.6: The effect of base concentration towards the conversion. Reaction conditions: 10 bar air, 80°C, 0.6 M of glycerol and catalyzed by Pd/Al$_2$O$_3$.

To investigate the importance of NaOH, the glycerol oxidation reaction was carried out at a neutral and three different basic concentrations, i.e. at 0.1, 0.5 and 0.8 M NaOH. The reactions at neutral conditions (not shown here) and at 0.1 M NaOH did not deliver any significant conversion (figure **3.1.6**), contrary to those carried out at higher NaOH concentration. Therefore, it could be reconfirmed that for this reaction basic condition was required in order to deliver a significant conversion.

3.1.6   Limitation to the reaction rate

It was well known that the O$_2$ mass transfer in the liquid phase and in the catalyst was not the limiting factor for the reaction rate as mentioned by several authors in this field.[120-121] More specific, it was mentioned that the mass transfer limitation can be omitted at stirring speed

45

higher than 550 rpm.[120] Therefore, for our system where the stirrer speed was 1500 rpm, the reaction rate was assumed to be limited only by the catalytic reaction.

3.1.7 Deactivation and reactivation of the catalysts

As previously mentioned, the conversion of this reaction could not reach higher values than approximately 50%. This might be caused by the deactivation of the heterogeneous catalysts. To prove this assumption, new fresh catalysts were introduced to the reactor after the conversion reached its maximum level.

Figure **3.1.7** shows that the conversion and the products yield did not change anymore after 10 hours of reaction, though the reaction runs for 25 hours. However, as the reaction had been stopped and new catalysts had been introduced, the conversion and the product yields increased. Hence, these results revealed that the limitation of the reaction is caused by the catalyst deactivation.

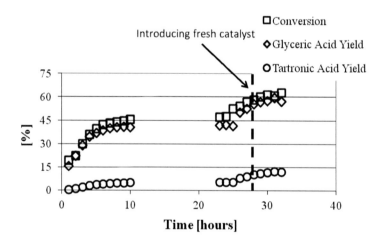

Figure 3.1.7: Conversion profile with additional new fresh catalysts. (The first reaction ran for 24 hours before the second reaction started and ran for 13 hours). Reaction conditions: 10 bar air, 80°C, 0.8 M of NaOH, catalyzed by Pd/Al$_2$O$_3$.

Since another source of catalysts deactivations might be originated from organic impurities, therefore, removing these impurities might improve the performance of the deactivated catalyst. Therefore, the catalyst recycling became an important issue for this reaction.

The recycling procedure consisted of two steps, firstly, the oxidation of organic impurities on the catalysts surface, secondly, the reduction of the oxidized catalysts to their metal form. In this experiment, the first step was carried out by heating the catalyst at 350°C under air and then nitrogen gas for three hours and the second steps was carried out by heating the catalyst at the same temperature under hydrogen environment for four hours. The similarity between the performance of the recycled and the fresh catalysts is shown in table **3.1.4**.

Table 3.1.4: The conversion profile of recycled and fresh catalysts.

	Conversion [%]	Yield [%] towards	
		Glyceric Acid	Tartronic Acid
Fresh Pd/Al$_2$O$_3$	42	34	7
Recycled Pd/Al$_2$O$_3$	42	34	4

Reaction conditions: 10 bar air, 80°C, 0.8 M of NaOH and 6 hours of reaction.

3.1.8 Glycerol oxidation reactions in a continuous stirred tank reactor

By applying the best reaction parameters based on the previous examinations, the catalytic aerobic glycerol oxidation reaction was carried out continuously. The experiment was conducted with a continuous stirred-tank reactor (figure **3.1.8**). A membrane pump was used to introduce liquid feed into the pressurized reactor. Additionally, the reactor was equipped with an external block heater. A more detail description of the system is presented in section **5.3.1**. Here, the study was focused on the effect of feed flow rates to the performance of the reaction. In the first three hours, the glycerol oxidation was carried out batchwise, then, glycerol was fed continuously at three different flow rates, i.e. 10, 20 and 30 ml/h. Simultaneous with the continuous feeding, the reactor was evacuated at the same flow rates.

Figure 3.1.8: A continuous stirred-tank reactor (series 4860 from Parr) with a membrane
pump.

Figure **3.1.9** presents the reaction conversion profile of the three different flow rates. At the
highest flow rate, the conversion declined just shortly after glycerol had been introduced
continuously. On the contrary, at the other lower flow rates the performance of the catalytic
process started to decline after seven and 12 hours of reaction depending on the flow rates.

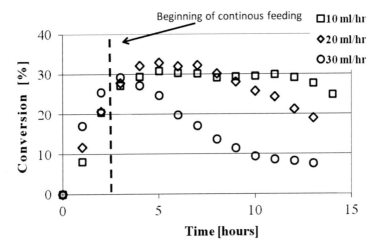

Figure 3.1.9: Conversion profile of continuous glycerol oxidation at different flow rates. (In the first three hours the reactions had been operated batch-wise before they started to run continuously for another 10-11 hours). Reaction conditions: 10 bar air, 80°C, 0.8 M of NaOH, catalyzed by Pd/Al_2O_3.

The capacity of the continuous reaction can be observed clearly in table **3.1.5** and figure **3.1.10**. At the medium and the lowest speeds the decline of glycerol conversion started after converting approx. 80 ml of additional feed which had been introduced continuously. Therefore, it could be concluded that the optimal flow rate was 20 ml/hour.

Table 3.1.5: Influence of flow rate to the amount of feed before activity decline. Reaction conditions: 10 bar air, 80°C, 0.8 M of NaOH, catalyzed by Pd/Al_2O_3.

Flow rate [ml/h]	Time before activity decline [h]	Amount of continuous feed before decline [ml]
10	8	80
20	4	80
30	-	-

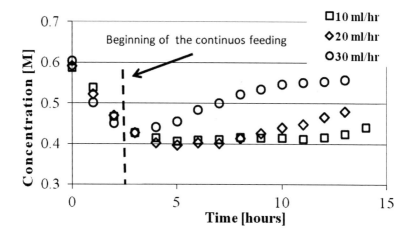

Figure 3.1.10: Concentration profile of glycerol during the continuous reaction. (In the first 3 hours the reactions had been operated batch-wise before they started to run continuously for another 10-11 hours). Reaction conditions: 10 bar air, 80°C, 0.8 M of NaOH, catalyzed by Pd/Al$_2$O$_3$.

3.2   Immobilized homogeneous catalysts for glycerol oxidation

The investigation of the immobilized system was initiated by choosing a cobalt-salen complex as a model of the homogeneous catalyst. The cobalt-salen complex was selected because of its capability to substitute precious metal catalysts in alcohol oxidation processes[122]. However, this catalyst is usually utilized as moveable molecules, thus making them able to roam freely in the reaction medium. To immobilize the catalyst molecules, the „Ship in a Bottle" encapsulation method via the zeolite MCM-22 was chosen.

The cobalt-salen was synthesized by reacting salicylaldehyde ethylenediamine (salen) with cobalt(II)acetylacetonate as shown in figure **3.2.1**.

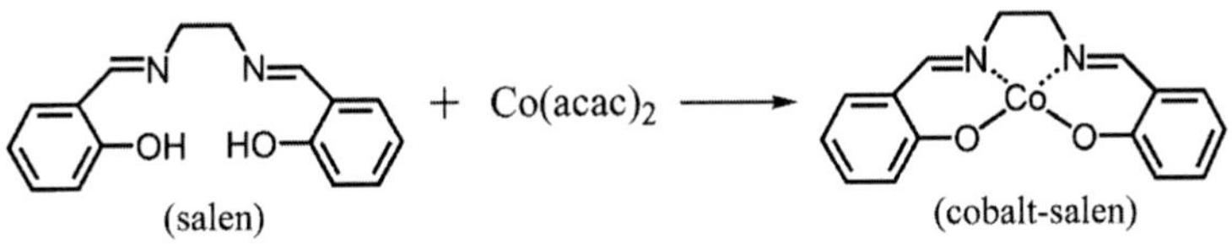

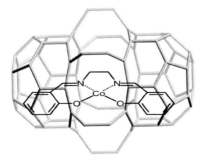

Figure 3.2.1: Synthesis of a cobalt-salen complex.

The cobalt-salen was immobilized by mixing with MCM-22 zeolite in ethanol solution. The immobilized complex is shown in figure **3.2.2**. The procedures of the synthesis of cobalt-salen complex and its immobilization in MCM-22 as well as the synthesis of MCM-22 are mentioned in sections **5.3.3 – 5.3.6**.

Figure 3.2.2: Cobalt-salen complex immobilized inside MCM-22 zeolite. The MCM-22 image was acquired from the Structure Commission of International Zeolite Association.[123]

The performance of the immobilization method was evaluated by observing the leaching value after a stability test. During this test the immobilized catalyst was exposed to a condition which is similar to the standard glycerol oxidation reaction, i.e. 60°C and 10 bar air in 0.3 M NaOH solution for 22 hours. After the experiment, via DLS analysis, cobalt leaching with a value of 0.1 ppm was detected.

Direct measurement via X-Ray diffraction (XRD) analysis revealed a significantly low cobalt content in the MCM-22 complex after the stability test. Table **3.2.1** notices about 60% decline of cobalt content after the test.

Transmission electron microscopy (TEM), which allows a more detailed observation of the MCM-22 surfaces, revealed slight inhomogeneous structures (see figure **3.2.3**) both before and after the stability test. The amorphous structures might be formed by zeolite ZSM-5, which is a typical byproduct from MCM-22 synthesis due to incomplete crystallization.[124]

Table 3.2.1: Element composition in the cobalt-salen-MCM-22 complex.

Elements	Composition [%]	
	before the stability test	after the stability test
Aluminum	9,45	11,26
Silicon	85,03	86,60
Cobalt	5,52	2,14
Total	100	100

Reaction condition: 60°C and 10 bar of air in 0.3 M NaOH solution for 22 hours.

The amorphous forms would be hindrance to substance diffusion[75] and likely trap the cobalt-salen complex inside their cage, thus preventing the absorption of the catalyst into the inner part of MCM-22 crystals. The non crystalloid parts, however, are prone to be invulnerable to a basic medium. Therefore, a substantial amount of them might drain away easily during the stability test while still holding the cobalt-salen complex.

(a)   (b)

Figure 3.2.3:   SEM figures of Zeolithe MCM-22 calcinated at 280°C. (a) Image of MCM-22 before stability test. (b) Image of MCM-22 after stability test.

Attempts were made to improve the crystal stability of the zeolite by increasing the crystallization temperature from 250 to 500°C, with 250°C being the standard temperature to produce salen-MCM-22 complexes. A better organized crystal structure was shown by the MCM-22 produced at the higher calcination temperatures (see figure **3.2.4**) than at lower temperatures (see figure **3.2.3**). This tendency was shown by the samples before and after the stability test alike.

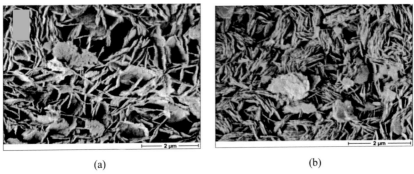

(a)   (b)

Figure 3.2.4:   SEM figures of Zeolithe MCM-22 calcinated at 500°C. (a) Image of MCM-22 before stability test. (b) Image of MCM-22 after stability test.

Unfortunately, a temperature stability test revealed that the cobalt-salen molecules were degraded at temperatures higher than 280°C. This result makes the implementation of this system not possible for the glycerol oxidation process.

3.3   A magnetic reactor system for glycerol oxidation reactions

3.3.1   Introduction

Due to the necessity to interrupt a continuous fixed bed catalytic glycerol oxidation reaction for catalyst regeneration, other reaction systems as alternatives were evaluated. Our initial assessment showed that a continuous slurry process system might enable the introduction and evacuation of catalysts without disrupting the reaction inside a reactor. A block flow diagram (BFD) displayed in figure **3.3.1** depicts the schematic graph of the slurry system.

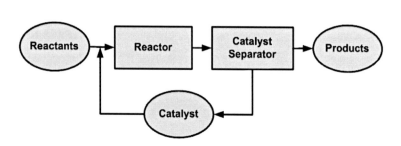

Figure 3.3.1:   A block flow diagram of a catalytic slurry system.

To facilitate an effective mass transfer and an optimum catalyst's volume to surface ratio, the slurry particles should be made as small as possible, e.g. smaller than 200 nm. However, the separation of very small particles such as Al_2O_3 and TiO_2 were challenging. Therefore, settling as a particle separation process was developed and modified to answer this challenge. Additionally, for the separation of particles that can be affected by a magnetic field, the utilization of magnetic separators offers an exceptional advantage to improve the separation performance. Therefore, in this section the development of the continuous glycerol oxidation process by exploiting the potential of magnetic field separations is described.

Because the common catalyst supports, such as activated carbon, TiO_2 and Al_2O_3 are not attracted by magnetic fields, it is necessary to substitute them with a magnetic material. Here, Fe_3O_4 was proven to be more suitable than γ-Fe_2O_3 (maghemite) due to its higher stability under alkaline condition in the presence of oxygen (see figure **A4.2** in the appendix **A4**).

The performance of catalysts such as gold and bimetal gold-platinum on the magnetic support for glycerol oxidation process will be mentioned in section **3.4**, while in this sub chapter only the performance of Fe_3O_4 as the sole catalyst will be elaborated. Moreover, this sub chapter also reports the development of the process including the selection of unit operations, their set up and the process performance in a continuous mode.

3.3.2   Glycerol oxidation in the presence of magnetic particles (Fe_3O_4)

Due to the harsh condition of the glycerol oxidation reaction such as an intensive contact with O_2 (10 bar air), high pH values, utilizing water medium at 100°C, etc. only the magnetic materials that can withstand these conditions can be chosen as the support. From the beginning of this experiment, it was found that a pretreatment of the Fe_3O_4 with glycerol or PEG prior to be used in glycerol oxidation reaction increases its stability under the harsh reaction condition.

Since Fe_3O_4 had not been known as a catalyst for glycerol oxidation reaction, the first test was carried out to examine its influence to the performance of glycerol oxidation reaction. By varying the weight ratio of glycerol to Fe_3O_4, it was shown that only with an excessive amount of the iron material, significant conversion can be observed (see figure **3.3.2**).

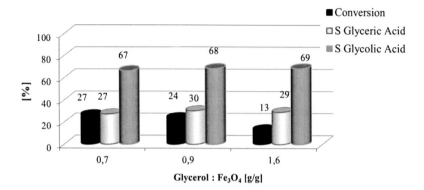

Figure 3.3.2: Conversion and selectivities of glycerol oxidation at different glycerol to Fe$_3$O$_4$ weight ratio. Reaction conditions: 10 bar air, 100°C, 0.15 M of NaOH, 4 hours of reaction.

Here, the ratio of 0.9 was chosen as the best option because lowering its value to 0.7 does not increase the glycerol conversion significantly. This value was, however, lower than the usual initial glycerol to metal catalyst weight ratios for the catalytic glycerol oxidation reactions, which are in the range between 500 and 3000.

After it had been proven that glycerol oxidation could be carried out in the presence of Fe$_3$O$_4$ without additional metal catalysts, the optimum reaction condition with respect to pressure and temperature was searched. In table **3.3.1** the performance of glycerol oxidation reaction at different pressures and temperatures are shown.

With the rise of the temperature the conversion increased from 24 to 64%. While at the temperature of 80 and 100°C the selectivity exhibited a high tendency towards glycolic acid, at 120°C the selectivity towards glyceric acid increased. The high tendency towards glycolic acid at low temperature was also observed in various glycerol catalytic reactions[40, 77], though this attribute usually appeared in a lower temperature range (less than 40°C). If glycolic acid was chosen as the target substance then the best reaction condition for the reaction would be 100°C of temperature and 10 bar air. Therefore, these values were selected as the standard reaction conditions for the next experiments.

Table 3.3.1: Effect of reaction pressure and temperature to the conversion and selectivity of glycerol oxidation in the presence of Fe_3O_4.

Temperature [°C]	Pressure [bar]	Conversion [%]	Selectivity (%)	
			Glyceric Acid	Glycolic Acid
80	10	24	30	64
100	10	44	27	71
120	10	64	45	52
100	5	18	14	80
100	15	14	27	72

Four hours of reaction. Initial glycerol and NaOH concentrations were 0.15 M.

Under the best reaction conditions which had been investigated for this experiment, the reusability of Fe_3O_4 was explored. After being separated the magnetic material was reused for a couple of times and the results exhibited only a minor discernible change between the fresh and the recycled Fe_3O_4 (see table **3.3.2**). This slight declining tendency of glycerol conversion was mainly caused by the Fe_3O_4 lost due to a practical limitation to recover the complete magnetic material as well as material dissolution represented by the leaching values.

Table 3.3.2:   Effect of reused Fe_3O_4 to the performance of glycerol oxidation.

Number of Recycle	Glycerol Conversion [%]	Leaching [ppm]
0	44	14
1	42	13
2	42	15
3	40	15
4	40	13

Four hours of reaction at 100°C and 10 bar air. Initial glycerol and NaOH concentrations were 0.15 M.

3.3.3   The set-up of the continuous magnetic slurry system

a.   Description of the miniplant

Based on the positive results of the batch experiments a miniplant (figure **A3.1** in appendix **A3**) to carry out a continuous operation of the magnetic slurry system was developed. In the following the development of the schematic graph depicted in figure **3.3.1** is described. Furthermore, a process flow diagram (PFD) displaying the important components of the miniplant is shown in figure **3.3.3**.

The miniplant is made up of two main parts, i.e. a reactor and a separator. The reactor comprises units such as an autoclave stirred tank reactor (CSTR) (figure **A3.2**), container (T1) holding the fresh as well as the recycled Fe_3O_4 solution, a vessel to supply the glycerol feed (T3) and a pressure releasing container (T2). The separator (figure **A3.3**) consists mainly of two magnetic separation containers (M1 and M2) and a final magnetic separator flask each equipped with an electromagnet. While the final liquid discharge from the separation part is collected in a product collector (T4), the slurry magnetic catalyst is pumped back to the reactor part via the recycled catalyst line. Furthermore, the miniplant is also equipped with several peristaltic and diaphragm pumps for transferring the pressurized and the unpressurized fluids, respectively. Finally, to monitor the material weight inside the containers from T1 to T4 each of them is equipped with a balance.

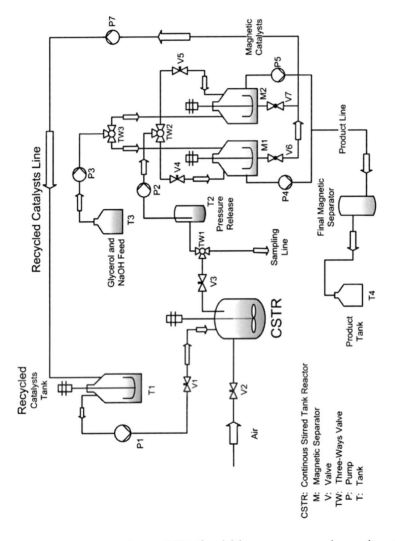

Figure 3.3.3: A process flow diagram (PFD) of a miniplant to carry out continuous glycerol oxidation reactions based on the magnetic slurry system.

b.   Process start up

A typical procedure to initiate the miniplant process is described in the following. Initially, the feed of glycerol-catalyst mixture was introduced from T1 into the CSTR where the reaction will take place. Furthermore, the temperature was elevated to 60°C. Next, the pressure inside the reactor was raised up to 5 bar by introducing air. Then the temperature was increased up to a target value (100°C) and additional air was introduced to increase the pressure up to a selected value (10 bar).

Inside the CSTR the reaction took place first batch-wise for several hours. The length of this step depended on the retention time setting. Afterwards glycerol-catalyst solution from T1 were introduced to the CSTR continuously. At the same time a comparable amount of solution inside the CSTR was evacuated and fed to the magnetic separation containers (M1 and M2) via the pressure releasing container.

c.   Continuous operation with magnetic recycle

The magnetic separation units started to operate after the first stream from the reactor (via the pressure releasing container) had arrived. During their operation these units worked batch-wise and alternately. At the beginning, the solution from the pressure releasing container (T2) had been introduced into one of the magnetic separation containers (M1) until it was half full. Then the flow was alternated to the other container (M2).

In the separation container M1, magnetic forces from the lower part of the flask settled the magnetic particles, thus separating them from the liquid phase. After the settling had been completed, the particle free liquid phase was evacuated from the container M1 by using the pump P4. This stream subsequently was introduced to the product tank T4.

The retained magnetic particles inside M1 were eventually mixed with a fresh glycerol solution which had been supplied from the vessel T3. Afterwards, this mixture was pumped into the recycled catalyst container T1 and then fed to the CSTR reactor. Accordingly, there would be no more new magnetic particles fed to the CSTR. At this point, the recycled system was started.

After the container M1 had been emptied, the flow of the product mixture from the CSTR (via pressure release container) was alternated from M2 back to M1. Then, the mixture in the container M2 followed the similar steps as in M1.

d.   Reaction conditions

The glycerol and Fe_3O_4 feeds used in this reaction were dissolved in a 0.15 M NaOH solution, while the initial glycerol concentration for this experiment was set at 0.15 M. Furthermore, the reaction temperature and pressure were fixed at 100°C and 10 bar air, respectively. Before running continuously, the reaction operated batch-wise with a residence time of 3 or 4 hours, while samples were taken every hour.

e.   Results of continuous experiments

The realization of the miniplant demonstrated successfully the implementation of the magnetic slurry concept as a continuous recycled oxidation process. Furthermore, it affirmed also the accuracy of equipment selection and setup for this process. Above all, by operating the miniplant steady-state performances of the continuous process were examined. In the following more details of the experimental results are discussed.

To anticipate a future comparison with the heterogeneous system discussed in section **3.1**, the target conversion of the continuous magnetic slurry system was set to 50 %, which was the maximum conversion of the former system. Under the optimal reaction condition as mentioned above, i.e. 100°C and 10 bar air, the reaction time to reach this value is four hours. The result of the first continuous experiment is presented in figure **3.3.4**.

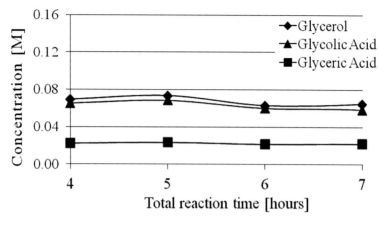

Figure 3.3.4: Substance concentration-time profiles of continuous glycerol oxidation reaction in the presence of Fe_3O_4 with 4 hours of residence time at 100°C and 10 bar air. Initial glycerol concentration was 0.15 M.

In the first continuous test, the concentration of glycolic and glyceric acids as well as glycerol did not display any significant fluctuation. Furthermore, their values showed only nominal deviations as presented in table **3.3.3**. Hence, it was concluded that the system is able to run steadily and the first continuous reaction isn't disturbed by any significant deactivations.

To see the influence of residence time on reaction performances a similar additional test was conducted with a shorter residence time, i.e. 3 hours. The result of the second continuous experiment is presented in figure **3.3.5** and their average values as well as their standard deviations are presented in table **3.3.3**.

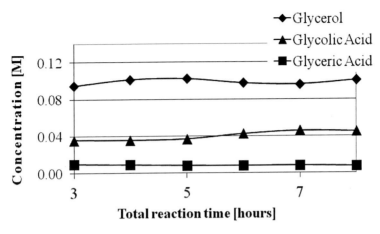

Figure 3.3.5:   Substance concentration-time profiles of continuous glycerol oxidation reaction in the presence of Fe_3O_4 with 3 hours of residence time at 100°C and 10 bar air. Initial glycerol concentration was 0.15 M.

As assumed, the product concentrations were lower with lowering the residence time. By shortening the residence time to 25%, the conversion was reduced from 50 to 30%. More importantly, the substance concentrations did not show any significant deviations. Therefore, it was proven that the miniplant can run steadily in different flow rates under the previously mentioned conditions.

In the third continuous experiment the recycled Fe_3O_4 from the first experiment was introduced. The substance concentration profiles from this test and their deviations can be seen in figure **3.3.6** and table **3.3.3**, respectively.

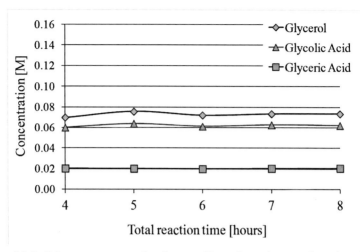

Figure 3.3.6: Substance concentration-time profiles of continuous glycerol oxidation reaction in the presence of recycled Fe_3O_4 with the residence time of 4 hours at 100°C and 10 bar air. Initial glycerol concentration was 0.15 M.

Table 3.3.3: The effect of recycled Fe_3O_4 on the performance of glycerol oxidation.

Residence time [h]	Fe_3O_4	Concentration [M]					
		Glycerol		Glyceric Acid		Glycolic Acid	
		$\bar{X}$	σ	$\bar{X}$	σ	$\bar{X}$	σ
3	Fresh	0.1	0.003	0.01	0.001	0.04	0.005
4	Fresh	0.07	0.005	0.02	0.001	0.06	0.007
4	Recycled	0.07	0.002	0.02	0.001	0.06	0.001

Reaction conditions: 100°C, 10 bar air and 0.15 M initial glycerol concentration.
$\bar{X}$: average value
σ: standard deviation

Based on comparative concentration profiles between figure **3.3.4** and **3.3.6**, the performance of the recycled magnetic slurry in the continuous experiment was proven to be similar with the fresh one. Consequently, this similarity elucidated also the feasibility of additional recycle steps for the continuous system. Finally, the success of these experiments can be used as a valid starting point for designing a continuous magnetic oxidation process at a more complicated and bigger scale.

3.4   Nano- and magnetic particles for glycerol oxidation reaction

3.4.1   Particles characterization

a.   Characterization of magnetic particles

One of the most critical aspects of the utilization of magnetic particles in oxidation reactions is their susceptibility against oxygen. Different observations of our magnetic particle, i.e., magnetite, such as visual observation, crystal structure analysis via an X-ray Diffractometer (XRD) and surface functional groups analysis via a Fourier Transform Infrared Spectroscopy (FTIR) showed that the magnetic particle was still stable after oxidation reactions. One example of the XRD spectrum of our magnetite is shown in figure **3.4.1**. There, we can observe the fingerprint peaks of the magnetite at 220, 311, 400, 422 and 511 nm^{-1} which are in accordance with the observations of Varanda and Yang[125-126].

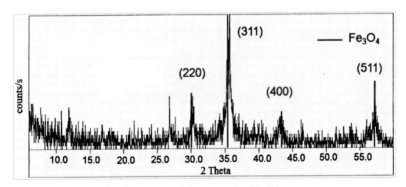

Figure 3.4.1:   X-ray diffraction pattern of the magnetite (Fe_3O_4).

Furthermore, figure **3.4.2** shows one example of the FTIR spectrum of our magnetite particles. This spectrum is similar to the one reported by the Csach's group[127]. Here, a noticeable peak at 571 cm^{-1} as the characteristic peaks of Fe-O can be observed, where a similar peak has been also observed by other authors at 575 cm^{-1}.[128-129]

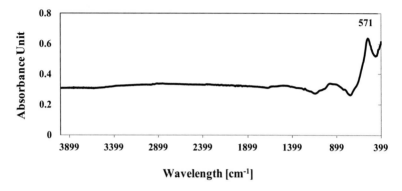

Figure 3.4.2: FTIR spectra of Fe$_3$O$_4$.

As mentioned in the preparation method part (section **5.3.7**), the magnetic particles Fe$_3$O$_4$ were functionalized by 3-aminopropyltriethoxysilane (APTES) to bind gold or other metal catalysts. Via an FTIR analysis, the presence of the APTES on the magnetic particles' surfaces was observed. The spectrum of the APTES functionalized magnetite is displayed in figure **3.4.3**, where noticeable characteristic peaks for Si-O at 1450, 1113 and 1050 cm^{-1} can be detected. Furthermore, the peak at 1650 cm^{-1} and the broad band around the peak at 3300 cm^{-1} as the characteristic peak for amine can also be noticed. Finally, the small peak at 2970 cm^{-1} as the sign of C-H stretching confirms the presence of the propyl group of APTES. These results are in accordance with the observation of Can[128] and Shaojun[129].

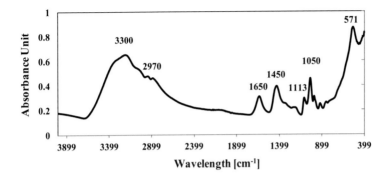

Figure 3.4.3: FTIR spectrum of Fe$_3$O$_4$ functionalized with APTES.

b.   The characterization of mono metal gold and platinum nanoparticles

The color changing of gold solution from transparent to red indicates the formation of metal nanoparticles (see figure **A4.1**). This observation was also supported by an Ultraviolet Visible Spectroscopy (UV-Vis) spectrum (figure **3.4.4**) which shows the changing in the maximum absorbance from 296 nm (the absorbance of the gold precursor) to 535 nm (the absorbance of the gold nanoparticles); thus, this indicates the production of gold nanoparticles[130].

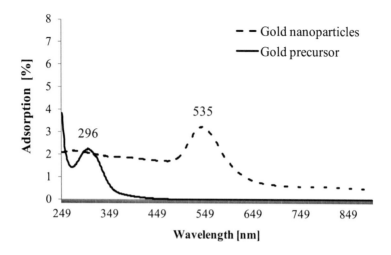

Figure 3.4.4: UV-Vis spectra of gold precursor (HAuCl₃) and gold nanoparticle solutions.

Moreover, an observation via transmission electron microscopy (TEM) confirmed also the formation of the gold nanoparticles. It showed a significant amount of particles having a diameter size of approx. 25-30 nm and having an icosahedron structure. Here, the image of gold nanoparticles produced via citric acid method is shown in figure **3.4.5**.

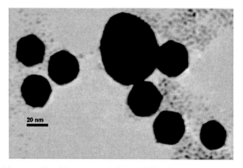

Figure 3.4.5: TEM image of gold nanoparticles stabilized by citric acid. The mole ratio between gold precursor (HAuCl₄.3H₂O) and citric acid is 1:4 (see section 5.3.8).

Further analysis via a Dynamic Light Scattering (DLS) equipment showed that 90% of the particles have a range of diameter between 20 to 50 nm with the mean value of 30 nm (see figure **3.4.6**). Thus, this observation confirmed the size measurement result of the TEM analysis. Accordingly, this size range still belongs to the optimum catalyst's particle size for the production of glyceric acid, where the size of the catalyst particles should be between 30 to 60 nm [26]. In addition, the gold nanoparticles showed a good stability, as it is shown in figure **3.4.6** that the size distribution of one month old gold particles was similar to the fresh sample.

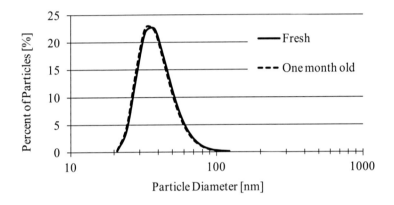

Figure 3.4.6: DLS measurement of particle size distribution of gold nanoparticles stabilized by citric acid.

When PVA was used as the stabilizing agent, instead of citric acid, smaller gold nanoparticles between 6 to 12 nm are formed as observed via DLS and TEM. The fact that PVA produces smaller nanoparticles is in accordance with other studies.[59]

Similar to the gold nanoparticles, the first observation of platinum nanoparticles production was carried via a UV-Vis instrument. In opposite to the gold nanoparticles, the platinum nanoparticles do not show any adsorption peak.[100] Therefore, the disappearance of the peak shown previously by the platinum precursor was used as the indication of nanoparticles formation (see figure **3.4.7**).

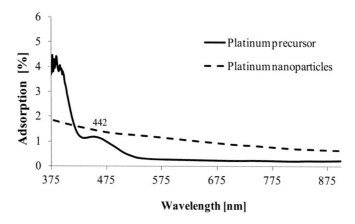

Figure 3.4.7: UV-Vis spectra of platinum precursor (H_2PtCl_6) and platinum nanoparticles solutions.

The histogram in figure **3.4.8** shows the particle size distribution of the platinum nanoparticles acquired by DLS analysis. Here, it was found that 99% of the platinum nanoparticles have a range of diameter between 16 to 50 nm, where the mean value is 26 nm. By comparing the particle diameters of gold and platinum nanoparticles, it is clear that the majority of platinum particles have a smaller diameter than the gold particles. Therefore, it could be assumed that a gold-platinum bimetal consists of gold as the core and platinum as the shell.

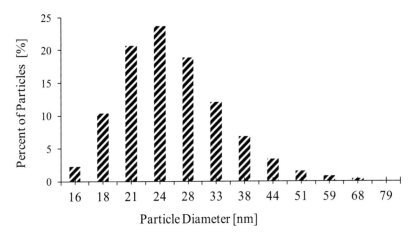

Figure 3.4.8: Particle size distribution of platinum nanoparticles observed via DLS.

c.   The characterization of bimetal gold-platinum nanoparticles

Similar to the platinum nanoparticles, for bimetal gold-platinum nanoparticles (the synthesis is described in section **5.3.8**), only the disappearance of the UV-Vis adsorption peak indicated the formation of the nanoparticles (see figure **3.4.9**). When these bimetal nanoparticles were analyzed via with DLS method, it was shown that 96% of the particles are in a range of diameter between 15 to 60 nm, with the mean value of 30 nm. Furthermore, this result was confirmed with the TEM analytic that shows uniform "flower-like" metal clusters having a diameter approx. 25-35 nm (see figure **3.4.10**).

Each metal cluster shown in figure **3.4.10** consists of many smaller platinum nanoparticles (approx. 3 nm) that arrange themselves around the bigger gold particles. An elemental analysis of the clusters carried out via Energy-dispersive X-ray spectroscopy (EDS) showed a significant presence of platinum and gold with a ratio of 1.3 to 1.

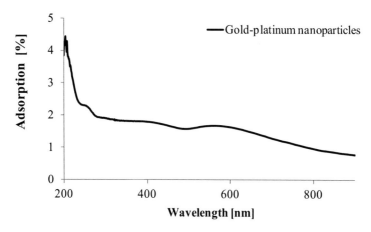

Figure 3.4.9:   UV-Vis spectrum of gold-platinum nanoparticles.

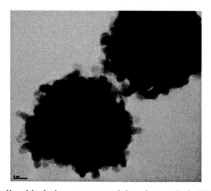

Figure 3.4.10:   Bimetall gold-platinum nanoparticles observed via TEM.

In figure **3.4.11**, the characteristic X-ray emissions of platinum and gold were measured along the line L1 that crosses over two bimetal clusters. The middle spectrum is the observed X-ray emission of platinum metals and the bottom spectrum is for the gold metals. These spectrums show that a high intensity of gold and platinum can be found in the middle of each cluster. However, in the vicinity areas of the clusters and in the gap between the two clusters the intensity of platinum metal is much higher than gold. Therefore, based on the EDS analysis the bimetal clusters show a gold core and platinum shell structure.

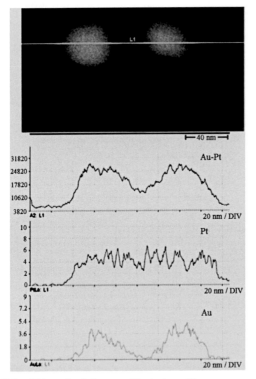

Figure 3.4.11: EDS of bimetal platinum gold clusters. The upper spectrum shows the cumulative signal of gold and platinum. The middle spectrum shows the signal of platinum. The bottom spectrum shows the spectrum of gold.

Furthermore, the presence of bimetal alloy was investigated by measuring the X-ray diffraction profile of the sample. Here, only the characteristic peaks of monometal gold, (which are highlighted) and platinum (the other peaks) can be identified and no sign of alloy can be observed (see figure **3.4.12**).

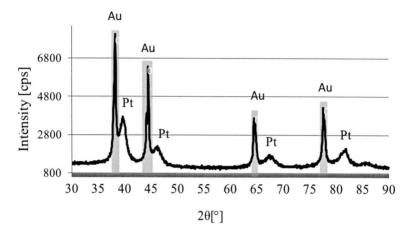

Figure 3.4.12: X-ray diffraction pattern of bimetal platinum gold clusters.

d.   The characterization of the growth of the bimetal gold-platinum nanoparticles

One important property of nanoparticles is their stability to keep their size at nanometer scale. Low stability causes a rapid agglomeration and precipitation of the metal particles. Thus, this condition is a disadvantage for most of catalytic reactions.

The observation of the bimetal nanoparticles' stability was carried out for several weeks. After one week aging, the size of the clusters showed no changing. However, the size of the two months old samples increased slightly with the time (observed via TEM). This result was also confirmed with the DLS observation of the mean size distribution of the clusters' particle diameter. It showed the particle diameters of 26 nm for the one week old sample and 35 nm for the two month old sample (see figure **3.4.13**).

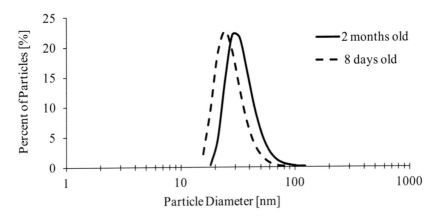

Figure 3.4.13:   Particle size distribution of the gold-platinum nanoparticles.

Besides the alteration of particle size, the composition of the clusters changed also with the time. An elemental analysis of the two month old clusters carried out via EDS showed a threefold increase of platinum content as compared to gold (see table **3.4.1**).

Table 3.4.1:   Platinum-gold ratio of the bimetal clusters.

Sample	Particle size d (nm)*	Mole-ratio**	
		Platinum	*Gold*
8 days old	26	1.3	1
2 months old	35	3.3	1

* Determined by DLS.
** Determined by elemental EDS.

An observation by TEM analytics showed the new formation of platinum shell layers on the surface of the clusters with the time. Interestingly, the new layers were formed from a platinum-platinum bound instead of platinum-gold. This fact can be seen clearly especially for the new samples. Although the surface of the gold particle was not completely covered, figure **3.4.14** shows that the new layers of platinum were built on the older layers instead on the unoccupied surface of the gold particle. Unfortunately, this phenomenon cannot be seen for older samples because of the very thick layer of platinum metal.

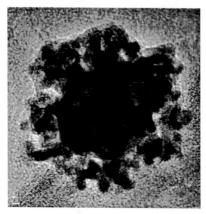

Figure 3.4.14:   8 days old bimetallic gold-platinum nanoparticles observed via TEM.

3.4.2   Glycerol oxidation catalyzed by magnetic catalysts

The experiment of glycerol oxidation was catalyzed by the magnetic mono- and bimetal catalysts (Au and Au-Pt on Fe_3O_4), where the metal content is 1% of the total catalyst weight. Several tests were carried out to investigate the effect of certain reaction parameters such as temperature, pressure, catalyst type and initial glycerol concentration on the performance of the catalytic reaction.

a.   Effect of temperature and reaction pressure as well the initial glycerol concentration

From the experiments in section **3.3**, it was found that the magnetic catalyzed glycerol oxidation reaction can be carried out optimally only at temperatures between 80 and 100°C due to the catalysts and products instability at higher temperatures. Furthermore, the magnetic support was very sensitive against oxidation at a pressure higher than 10 bar of air. Therefore, the investigation of pressure and temperature effects was carried out only in a small experimental window, i.e. at temperatures of 80 and 100°C and at pressures from 6 to 10 bar.

For this purpose, the initial glycerol concentration was set at 0.14 M and the results of this investigation are shown in table **3.4.2**.

Table 3.4.2: Performance of magnetic gold and gold-platinum catalysts for glycerol oxidation at different pressures and temperatures.

Catalyst	Pressure [bar]	Temp. [°C]	Conversion [%]	Selectivity (%)			
				Glyceric Acid	Glycolic Acid	Tartronic Acid	Formic Acid
Au/Fe$_3$O$_4$	6	80	8	13	62	n.d.	20
	6	100	44	11	68	n.d.	19
	10	100	98	9	64	n.d.	24
Au-Pt/Fe$_3$O$_4$	6	80	55	64	13	13	8
	10	100	93	36	46	9	8

n.d.: not detected.
The initial glycerol concentration was 0.14 mol/l.

At 80°C the conversion of reactions catalyzed by Au-Pt/Fe$_3$O$_4$ was higher than the conversion of the reaction catalyzed by Au/Fe$_3$O$_4$. The superiority of Au-Pt over Au catalysts was in accordance to the work of Prati's group which used carbon support material.[44] On the contrary, at 100°C the activity of the gold catalyst was slightly higher than the gold-platinum catalyst.

The type of catalyst determined also the selectivity of the reaction. Table **3.4.2** shows that the selectivity of Au/Fe$_3$O$_4$ is significantly higher towards the production of glycolic acid; hence, this indicates the preferable oxidation of the secondary alcohol group of glycerol. On the contrary, the Au-Pt/Fe$_3$O$_4$ showed only a slightly higher selectivity toward glycolic acid than to glyceric acid at 100°C. Furthermore, the reaction catalyzed by this catalyst showed a significant selectivity towards tartronic acid.

From our previous experiment we found that a high glycerol concentration limited the reaction conversion. On the other hand, the initial reactant's concentration should be set as

77

high as possible to reduce the reactor investment cost, i.e. increasing the space-time yield. By increasing the initial glycerol concentration from 0.14 to 0.71 and 1.41 M, it was shown that 0.71 M was the optimum concentration with respect to the conversions and the amount of glycerol which has been converted (see table **3.4.3**).

Table 3.4.3: Performance of magnetic gold and gold-platinum catalysts for glycerol oxidation at different glycerol initial concentration.

Catalyst	Glycerol conc. [mol/l]		Conversion [%]	Selectivity (%)			
	Initial	Converted		Glyceric Acid	Glycolic Acid	Tartronic Acid	Formic Acid
Au/Fe$_3$O$_4$	0.14	0.14	98	9	64	n.d.	24
	0.71	**0.31**	**43**	59	26	n.d.	12
	1.41	0.32	24	43	41	n.d.	15
Au-Pt/Fe$_3$O$_4$	0.14	0.13	93	36	46	9	8
	0.71	**0.2**	**38**	61	21	8	8
	1,41	0.21	15	60	34	1	5

n.d.: not detected.
Reactions were carried out at 100°C and 10 bar air.

As the initial glycerol concentration increased from 0.14 to 0.71 M, the selectivity towards glycolic acid decreases and the selectivity towards glyceric acid increases. However, at 1.41 M of initial concentration the selectivity towards glycolic acids rose until almost reaching the same value as the selectivity to glyceric acid for catalyst Au/Fe$_3$O$_4$. Additionally, a further oxidation product of glyceric acid, i.e. tartronic acid could be observed from the Au-Pt/Fe$_3$O$_4$ catalyzed reaction especially at lower initial concentration.

Because the reutilization of catalysts is one of the most important aspects in catalytic processes, we investigated the reuse of our magnetic catalysts. After they had been reused for one time, the magnetic catalysts showed an activity alteration. For the catalyst Au/Fe$_3$O$_4$, the conversion decreased from 43 to 34% and for catalyst Au-Pt/Fe$_3$O$_4$ the conversion decreased slightly from 38 to 35%. The product selectivity changed slightly (see table **3.4.4**). Furthermore, from all residues no catalyst leaching could be observed via inductively coupled plasma (ICP) analyses.

Table 3.4.4:   Performance of recycled catalysts.

Catalyst	Catalyst Status	Conversion [%]	Selectivity (%)			
			Glyceric Acid	Glycolic Acid	Tartronic Acid	Formic Acid
Au/Fe$_3$O$_4$	Fresh	43	61	26	n.d.	11
	Reused	34	37	52	n.d.	10
Au-Pt/Fe$_3$O$_4$	Fresh	38	61	21	8	8
	Reused	35	56	28	8	7

n.d.: not detected.
Reactions were carried out at 100°C and 10 bar air. The initial glycerol concentration was 0.7 M.

3.5   Economic aspects of the magnetic catalytic system for glycerol oxidation reactions

This subchapter presents a brief material cost-revenue analysis of the glycerol oxidation reaction based on the magnetic process catalyzed by Au/Fe$_3$O$_4$. Here, the production of 10,000 ton/a glyceric acid was selected as the basis of the process. Therefore, the reaction should run under optimal conditions for producing glyceric acid as described in section **3.4**. Besides producing glyceric acid, this process produces also glycolic and formic acids. Under these conditions glyceric, glycolic and formic acid could be produced with yields of 26, 11 and 4%, respectively (see the performance of the fresh Au/Fe$_3$O$_4$ depicted in table **3.4.4**). The price of glyceric acid was set similar to lactic acid, another C3 acid, i.e. at 1,200 €/ton. Furthermore, the price of glycolic acid was priced similar to acetic acid, i.e. 680 €/ton.[131]

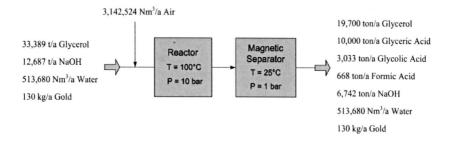

3,142,524 Nm³/a Air

33,389 t/a Glycerol
12,687 t/a NaOH
513,680 Nm³/a Water
130 kg/a Gold

Reactor
T = 100°C
P = 10 bar

Magnetic
Separator
T = 25°C
P = 1 bar

19,700 ton/a Glycerol
10,000 ton/a Glyceric Acid
3,033 ton/a Glycolic Acid
668 ton/a Formic Acid
6,742 ton/a NaOH
513,680 Nm³/a Water
130 kg/a Gold

Figure 3.5.1: Block flow diagram showing an overview of the material balance of magnetic slurry glycerol oxidation.

Referring to the block flow diagram (BFD) displayed in figure **3.5.1**, the production of 10,000 tons of glyceric acid requires 33,389 tons of glycerol. Moreover, 3,033 tons of glycolic and 668 tons of formic acids would also be produced from this process. Nevertheless, the flow diagram shows only the simplest possible alternative of industrial application for this system. Hence, neither recycling steps nor energy cost had been taken into consideration. Based on this diagram and considering only the disparity between the raw material costs and the product prices the profit of the industrial scale glycerol oxidation was evaluated (see table **3.5.1**).

As depicted in table **3.5.1**, the financial calculations of the flow diagram resulted in a negative value, i.e. a loss of € 10,630,363 annually. Therefore, several modifications were made in order to evaluate the financial feasibility of the magnetic system in general. The first modification was the introduction of a water recycling step. Thus, the cost of water could be reduced. Unfortunately, the final cost might increase due to the additional energy cost for this step (see option 2 in table **3.5.2**). Another alternative was the prolonged utilization of the gold catalyst. However, extrapolating the operation time of the gold catalyst by six folds (a same practice made by Katryniok group[132]) still did not generate a positive balance sheet, i.e. a loss of € 10,449,318 annually.

Table 3.5.1: Materials' cost and revenue of industrial scale glycerol oxidation reaction.

Items	Price/Unit	Amount p.a. Produced	Amount p.a. Consumed	Cost and Revenue [€ p.a.]
Glycerol	555 €/ton[133]		33,389 ton	- 18,530,989
Glyceric Acid	1,200 €/ton[a]	10,000 ton		12,000,000
Glycolic acid	680 €/ton[b]	3,033 ton		2,062,500
Formic Acid	80 €/ton[134]	668 ton		53,405
NaOH	440 €/ton[131]		12,687 ton	- 5,582,198
Air	1.79 €/100 Nm³[c]		3,142,524 Nm³	- 56,251
Water	0.7 €/m³[c]		518 Nm³	-359,576
Gold Recycling	1,673 €/kg[d]		130 kg	- 217,254
Profit (or loss)				-10,630,363

[a] Glyceric acid price is set similar to lactic acid. Lactic acid price is 1,200€/ton.[131]
[b] Glycolic acid price is set similar to acetic acid. Acetic acid price is 680€/ton.[131]
[c] Source Emery Oleochemicals.
[d] Source Umicore.

Hypothetically omitting the glycerol cost might be justified since this dissertation was partially motivated by a possible boom of the biodiesel industry, wherein the price of glycerol may decline to zero value. This alternative gives a positive value in the economic balance i.e. a profit of € 7,900,626 annually (option **4** in table **3.5.2**). By prolonging the operation time of the gold catalyst by six folds, the revenue of this process can be improved (option **5** in table **3.5.2**).

Charging glyceric acid at the recent price may deliver a significant profit (see option **6** in table **3.5.2**). Unfortunately, these prices were too high for a bulk chemical price. As a comparison, the price of another bulk chemical such as lactic acid is only 1,200 €/ton. Fortunately, although by charging the glyceric acid at 1% of the recent price, this process can still give a significant profit. Therefore, from the financial point of view, the slurry magnetic glycerol oxidation process may be profitable if the price of glycerol decreases significantly.

Table 3.5.2: Alternative scenarios of the industrial scale glycerol oxidation process producing 10,000 ton/a of glyceric acid.

Nr.	Options	Profit or loss [€ p.a.]
	Product as Bulk Chemical	
1	Original design	-10,630,363
2	Original design + water recycling	-12,777,291
3	Original design but the gold reused for 6 times	-10,449,318
4	Original design with no charge for glycerol	7,900,626
5	Combination of options **3** and **4**	8,081,671
	Product not as Bulk Chemical	
6	Original design but selling the glyceric acid at the recent price*	2.3 billions
7	Original design but the glyceric acid should be sold at 1% of the recent price* ($\approx$ 2,334 €/ton)	713,110
8	Option **4** but the glyceric acid should be sold at 1% of the recent price* ($\approx$ 2,334 €/ton)	19,244,099

* The recent price of glyceric acid: 23.3 €/100gr.[135]

3.6 Glycerol for an environmental friendly synthesis of nanoparticles

3.6.1 Introduction

The synthesis of metal nanoparticles via sol-gel method in the preparation of heterogeneous catalysts for the catalytic glycerol oxidation process were usually carried out by utilizing harmful reducing and stabilizing agent materials such as hydrazine, sodium borohydride, etc. (see sections **2.2.8** and **2.4.2.1**). Due to the increase of health and environmental awareness in chemical industries any attempt to find a substitution of these harmful materials is highly desirable.

In an aerobic glycerol oxidation, glycerol is oxidized while in the same time the oxidation agent, molecular oxygen, is reduced in the present of metal catalysts. Based on the reducing capability of glycerol, it was assumed that when this reaction is carried out without the presence of oxygen, glycerol will be able to reduce certain metal ions precursor such as $PdCl_2$, $PtCl_2$, $AuCl_3$, etc. (see section **2.4.2.1**). Herein, the synthesis of nanoparticles from certain metal ion precursors by utilizing glycerol as reducing and stabilizing agent is presented.

3.6.2   Formation of nanoparticles

The color change of a metal precursor solution might serve as an initial indicator of the formation of nanoparticles. For example, here the solutions' color change from orange gold to red ruby for gold or from olive green to dark brown for cobalt. Furthermore, for some metal nanoparticles the changing in their characteristic UV-Vis spectra can also indicate the formation of nanoparticles from their ionic precursor, for example, the shifting of the maximum adsorption from 297 to 532 nm for gold (figure **3.6.1**) or the disappearing of the maximum adsorption in case of cobalt (figure **3.6.2**). Considering their simplicity, these analytic methods were used in the initial experiments for finding the optimum reaction conditions to produce metal nanoparticles.

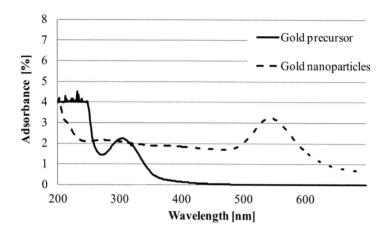

Figure 3.6.1: UV-Vis spectra of gold precursor and nanoparticles.

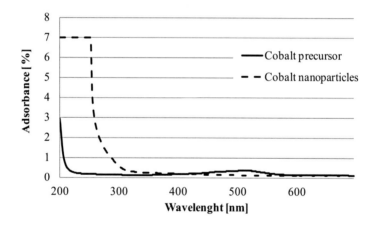

Figure 3.6.2: UV-Vis spectra of cobalt precursor and nanoparticles.

The investigation revealed that the synthesis of metal nanoparticles based on glycerol method should be carried out at temperature between 40 to 80°C. In this range, the nanoparticles were formed after four to 12 hours depending on the reaction temperature. The optimum glycerol to metal precursor molar ratio was 65:1 for precursors such as $RuCl_3$, $HAuCl_4$, H_2PtCl_6, and $Pd(OAc)_2$. On the other hand, due to its less reactivity, the synthesis of cobalt nanoparticles required higher amount of glycerol (150 mol glycerol for one mol of $CoCl_2$). The maximum ratio of glycerol to metal precursors allowed before the nanoparticles start to precipitate is 750:1. In addition, NaOH was added to improve the reducing capability of the glycerol since an alkali source is also well known in supporting the reduction of metal ions.[97-98]

3.6.3   Morphology of the nanoparticles

Based on the success of the initial test, a transmission electron microscopy (TEM) unit was utilized to investigate the size and the morphology of the nanoparticles. By utilizing glycerol as reduction and stabilizing agents, nanoparticles of certain noble metals with a size smaller than 10 nm were produced. The nanoparticles of ruthenium (figure **3.6.3a**), palladium (figure **3.6.3b**) and platinum (figure **3.6.5a**) formed individual small particles and were well distributed. On the other hand, the gold nanoparticles formed agglomerations which were distributed poorly (figure **3.6.3c**). As shown by the TEM image, the gold nanoparticles were stacked on each others.

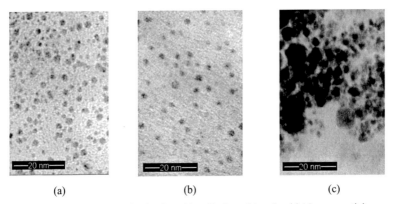

(a)   (b)   (c)

Figure 3.6.3: TEM pictures of ruthenium (a), palladium (b) and gold (c) nanoparticles synthesized in the presence of glycerol.

By comparing the particle sizes observed via TEM and Dynamic Light Scattering (DLS) analysis, slightly different results were revealed between both measurements (table **3.6.1**). These differences were due to the characteristics of DLS analysis which also includes the hydrodynamic surrounding (mainly glycerol) of a particle in the size measurement. In addition, for gold nanoparticles the DLS analysis showed only the size of the aggregates instead of the size of each single particle (DLS analysis indicated the presence of particles having sizes of 60 nm).

Table 3.6.1:   Comparison of nanoparticle diameters measured via TEM and DLS

Metal	Particle diameter measured via-	
	TEM (nm)	DLS (nm)
Ruthenium	2	10
Platinum	1-5	12
Cobalt	7	10
Gold	3-18 (stacked)	60

Colloid stability analysis confirmed that the platinum, ruthenium and cobalt nanoparticles were stable even for more than two months. Although in longer time, a slight increase in particle size was noticed though this did not lead to precipitations. On the other hand, gold nanoparticles started to precipitate even after just one week. This showed that gold nanoparticles require stronger stabilizing agents such as electrostatic protection of anionic ligands[136], [59] or steric protection of polymers[80]. These facts were also in agreement with other authors where an addition of a stabilizing agent was required for producing gold nanoparticles via polyol based methods[85], while for the synthesis of platinum nanoparticles only the presence of an alcohol was sufficient[94].

3.6.4   Comparison with other reduction agents

After the observation of the particle size and morphologies of the nanoparticles, the performance of glycerol as a reduction agent was compared with other reduction agents, such as $NaBH_4$ and H_2. For this purpose, we chose $Co(OAc)_2$ as our model precursor.

Transmission Electron microscopy images (figure **3.6.4**) showed that cobalt nanoparticles which were synthesized by utilizing glycerol had particle diameters smaller than 10 nm (figure **3.6.4b**). They were slightly smaller than the nanoparticles reduced by $NaBH_4$ (figure **3.6.4a**) and significantly smaller compared to nanoparticles reduced by H_2 (figure **3.6.4c**). Nevertheless, all cobalt nanoparticles formed spherical shapes.

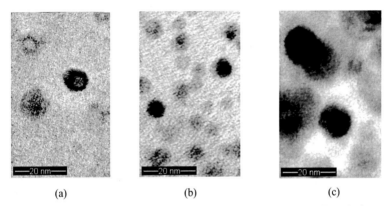

(a)   (b)   (c)

Figure 3.6.4: TEM Images of Cobalt Nanoparticles which were produced by reducing $Co(OAc)_2$ with (a) $NaBH_4$, (b) Glycerol and (c) H_2.

To prove the statement that glycerol played an important role in the synthesis of nanoparticles, blank experiments were made where all reaction conditions are identical except the presence of glycerol. As represented in figure **3.6.5**, the TEM image of platinum nanoparticles which was synthesized in the presence of glycerol showed well distributed formation of nanoparticles. On the contrary, only a few of individual bigger particles were formed for the sample which was synthesized without glycerol. Moreover, this formation has taken place only during the TEM analysis due to the activation by electron radiation. Here, the contribution of glycerol as reduction agent in the synthesis of nanoparticles could be proven.

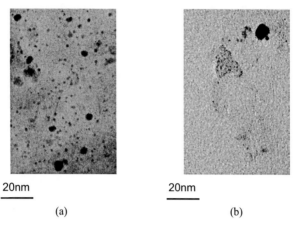

20nm   20nm

(a)   (b)

Figure 3.6.5: TEM photos of platinum nanoparticles synthesized in the presence of glycerol (a) and in the absence of glycerol (b).

4   CONCLUSIONS

The first objective of this dissertation was attained by examining the performance of commercial catalysts for fixed bed glycerol oxidation reactions. A slight superiority of palladium over platinum especially regarding the resistance against catalyst deactivation was reaffirmed. Moreover, an advantage in mechanical stability of using Al_2O_3 pellets rather than graphite granulates as metal catalyst supports was revealed. Furthermore, the Pd/ Al_2O_3 catalyst led to the production of glyceric acid with a maximum yield of 34 %, while generating tartronic acid as a byproduct with a yield of 8 %.

This examination also disclosed the **requirement of oxygen pressure higher than 1 bar**, irrespective of the amount and the method of oxygen supply, as shown by the trickle bed and fluidized bed experiments. The experiments proved the exclusive role of the dissolved oxygen in the aerobic catalytic reaction, taking into account that at an equilibrium state the amount of gas in a liquid phase depends only on the pressure at a constant temperature. Consequently, this result refuted any postulate of direct oxygen consumption from the gaseous phase.

The limited conversion of glycerol, which was only up to 50%, led to the investigation of factors that might confine the reaction. It was assumed that the decline was caused by the deactivation of the catalysts. This was proven by the fact that changing the spent catalyst with fresh catalyst enabled a further conversion. Finally, attempts to recover the catalyst into its original condition were successfully executed via a thermal treatment in hydrogen and nitrogen atmospheres.

Conducting the reaction continuously reaffirmed catalyst deactivation as the main process obstacle. The deactivation was caused by impurities that can only accumulate on the surface of the fixed bed catalyst but not in the liquid medium, considering the steady flow through the reactor. Moreover, the continuous experiments also highlighted an **inherent weakness of the fixed bed catalyst application**, i.e. the inevitable interruption of the process for reactivating the catalyst. Furthermore, considering the rapid activity decline initiated the pursuit of alternative systems that enable catalyst regeneration without interrupting the continuous process.

Besides developing new systems as the follow-up of the first continuous operation test, in order to reach its second objective, this dissertation evaluated a synthesis of a novel catalyst produced by the **immobilization of a cobalt-salen homogeneous catalyst in MCM-22 zeolites**. The cobalt-salen was chosen due to its favorable prospects for catalytic alcohol oxidation. Furthermore, the encapsulation with the MCM-22 was considered as one of the best immobilization methods, while it will not alter the structure of the homogeneous catalyst, contrary to covalent binding.

Leaching of cobalt into the reaction medium was an important parameter in the evaluation of the immobilization. However, the measurements conducted by ICP-EOS and X-Ray diffraction (XRD) exposed two contradicting results. Whilst no leaching could be detected via the ICP analysis, the XRD revealed about 60% of cobalt leaching after a stability test. This contradiction might be caused by the failure of the ICP analysis to detect amorphous zeolites, which were likely formed due to imperfect crystallization.

The NaOH-sensitive amorphous structures located in the periphery of the MCM-22 were prone to contain more cobalt-salen catalysts than the inner zeolite crystals. Therefore, when the former were wiped out during the stability test, they would drag the catalysts to the liquid medium, thus reducing the cobalt content in the remaining MCM-22 zeolite.

Increasing the zeolite synthesis temperature up to 500°C enhanced the quality of crystallization, thus reducing the formation of amorphous structures. However, since the cobalt-salen was especially susceptible to degradation at temperatures higher than 280°C, the high temperature synthesis of the MCM-22 was omitted, thus dismissing the application of the immobilization method.

In a next step, an alternative continuous system which is independent from catalyst regeneration, the problem of the fixed bed catalyst system, was elaborated. Evaluating the settling process of CeO_2, TiO_2 and Al_2O_3 particles, having a size less than 200 nm, highlighted the poor performance of the regular methods, where a complete process might take more than 20 hours. Fortunately, the **settling time could be reduced** significantly to 30 minutes **by applying magnetic fields.** For that purpose, the utilization of magnetic particles was necessary. Therefore, this was followed by the comparison of magnetic materials, which then emphasized the superiority of Fe_3O_4 over γ-Fe_2O_3 due to the former stability in basic solutions in a pressurized oxygen atmosphere.

Along with its role as a support material, the **Fe_3O_4 also showed a catalytic capability** to deliver a significant conversion of glycerol to glyceric acid with a yield of 31%. Unfortunately, due to the sensitivity of the magnetic material, the reaction conditions were limited to a maximum temperature and pressure of 100°C and 10 bar, respectively.

Taking advantage of the Fe_3O_4 converting ability, a **miniplant** was built **for realizing the continuous magnetic glycerol oxidation.** While the conversion of the continuous system was slightly higher than the batch-wise system, i.e. 53 and 44%, respectively, the selectivity towards glyceric and glycolic acids in both systems were similar. More interestingly, the magnetically recycled Fe_3O_4 delivered a similar performance to the fresh Fe_3O_4. This validated the feasibility of the magnetic particles to replace the fixed bed system for aerobic catalytic glycerol oxidation, thus achieving the third objective of the dissertation.

As the follow-up, the fourth objective of the dissertation was attained after the **magnetic gold and gold-platinum catalysts** for the aerobic glycerol oxidation had **been developed.** Here, the formation of Fe_3O_4 magnetic crystals was validated via the XRD and Fourier Transform Infrared Spectroscopy (FTIR). Furthermore, the formation of gold and platinum nanoparticles was detected via Ultraviolet Visible Spectroscopy (UV-VIS) by observing the alteration of the maximum absorbance wavelengths.

The morphologies of the gold and gold-platinum catalysts were unveiled via a Transmission Electron Microscope (TEM). The particles displayed icosahedron forms of the gold metal and flower-like forms of the gold-platinum bimetals. A reexamination of the morphologies after two months of storage indicated a slight enlargement of the bimetals. In contrast, the size of gold metal remained unchanged.

A detailed size measurement of the bimetal particles was conducted with a Dynamic Light Scattering (DLS) instrument. The observation showed a mean value of 26 and 30 nm of the fresh and the two months old samples, respectively. Furthermore, an elemental analysis of the bimetal clusters carried-out via Energy-dispersive X-ray spectroscopy (EDX) confirmed the gold core and platinum shell formations with a ratio of 1 to 1.3 and 3.3 for the new and the older samples, respectively. Therefore, these findings proved that the enlargement was originated by the growth of the platinum shell.

The **selectivity** of both catalysts **depended** highly on the **initial glycerol concentration**. At lower concentrations the catalyst Au/Fe_3O_4 had a higher tendency towards the oxidation of glycerol's secondary alcohol (producing glycolic acid) while at higher concentrations a tendency towards the primary alcohol of glycerol (producing glyceric acids) rose. A similar tendency was also shown by the catalyst $Au-Pt/Fe_3O_4$. Moreover, the latter catalyst showed a high activity especially at low temperatures and an ability to produce tartronic acid. Preferring the production of glyceric acid, at the optimum reaction conditions the yield was 26%.

An economic analysis based on the glycerol oxidation catalyzed by the Au/Fe_3O_4 was discussed. With a target production of 10,000 ton/a glyceric acid, the first calculation of the material costs and the product prices unfortunately resulted in a negative value. Evaluation of different scenarios, such as omitting the cost of glycerol, can fortunately exhibit a significant potential profit. Therefore, it could be concluded that the continuous magnetic glycerol oxidation process might be feasible if the price of glycerol decreases significantly.

Finally, the last objective of this dissertation was achieved after the synthesis of certain nanoparticles from metal ion precursors by **utilizing glycerol as a reduction agent** had been described. Except for gold and iron, the nanoparticles produced showed a good stability for a long time. Moreover, the glycerol-based method could produce smaller nanoparticles compared to the NaBH₄ or the hydrogen reduction methods. Nevertheless, especially for cobalt nanoparticles, their morphology showed spherical shapes.

A slight difference in particle size measurements between the TEM and DLS analytics might indicate the role of glycerol in covering the nanoparticles. This proved the role of **glycerol as a stabilizing agent** in addition to a reduction agent. Furthermore, the glycerol-based method could be carried-out at lower temperatures compared to most other polyol-based methods and requires no additional harmful reduction and stabilizing agents. Therefore, this synthesis method might become one of the best environmental friendly methods for the production of nanoparticles.

5 EXPERIMENTAL PART

5.1 Chemicals

Table 5.1.1: Material list

Substances	Purity	Source
(3-Amino propyl)triethoxysilane (APTES)	99%	Acros Organics (Nidderau, Germany)
Ammonium hydroxide (NH_4OH)	25%	Fluka (Buchs, Switzerland)
Ascorbic acid	99%	Across Organics (Nidderau, Germany)
Citric acid	99%	Across Organics (Nidderau, Germany)
$Co(acac)_2$	99%	Across Organics (Nidderau, Germany)
Glycerol	99.5%	Emery Oleochemicals (Düsseldorf, Germany)
Glyceric acid	99%	Sigma-Aldrich Chemie (Taufkirchen, Germany)
Glycolic acid	99%	Sigma-Aldrich Chemie (Taufkirchen, Germany)
Gold (III) chloride trihydate ($HAuCl_4 \cdot 3H_2O$)	50,16%	Umicore (Hanau, Germany)
Hexachloridoplatin acid ($H_2PtCl_6 \cdot 6H_2O$)	40%	Umicore (Hanau, Germany)
Hydrogen	$\geq 99.9\%$	Air Liquide (Dortmund, Germany)
Hydroxypyruvic acid	$\geq 95\%$	Fisher Scientific (Nidderau, Germany)
Iron (III) chloride hexahydrate ($FeCl_3 \cdot 6H_2O$)	99%	Across Organics (Nidderau, Germany)
Iron (II) chloride tetrahydrate ($FeCl_2 \cdot 4 H_2O$)	99%	Merck (Darmstadt, Germany)
Nitrogen	$\geq 99.9\%$	Air Liquide (Dortmund, Germany)
Oleic acid	Extra pure	Fisher Scientific (Nidderau, Germany)

Substances	Purity	Source
Oxalic acid	≥99%	Fisher Scientific (Nidderau, Germany)
Oxygen	≥ 99.9%	Air Liquide (Dortmund, Germany)
Palladium on activated carbon	0.5% (Pd)	Degussa (Essen, Germany)
Palladium on Al_2O_3	0.5% (Pd)	Degussa (Essen, Germany)
Platinum on Al_2O_3	0.5% (Pt)	Degussa (Essen, Germany)
Polyethylene glycol /4000 G	Extra pure	Across Organics (Nidderau, Germany)
Polyvinyl alcohol	99%	Across Organics (Nidderau, Germany)
Sodium borohydride ($NaBH_4$)	≥ 98%	Merck (Darmstadt, Germany)
Sodium hydroxide (NaOH)	≥ 99%	Merck (Darmstadt, Germany)
Synthetic air (20.5% O_2)	≥ 99.9%	Air Liquide (Dortmund, Germany)
Tartronic acid	98 %	Alfa Aesar (Karlsruhe, Germany)
Water	Distilled	Own production

Other chemicals were obtained from Acros Organics.

5.2   Analytic equipments and methods

5.2.1   High performance liquid chromatography (HPLC)

The quantitative analysis of glycerol oxidation reaction was carried out via a LaChrome Elite HPLC instrument from Hitachi (Tokyo, Japan). The instrument was equipped with an auto-sampler L2200 and an oven L2350 from the same company.

The liquid samples were separated via a Trentec 308R-Gel.H chromatography column from Trentec Analysentechnik (Gerlingen, Germany). Standard substances such as glyceric, tartronic, oxalic, glycolic and formic acids were used to identify the reaction products by comparing the retention time. Additionally, the peak areas of these substances were related to their concentration, where by this relation the product concentration was determined. The quantitative analyses used external calibrations. The detail of the analytic conditions is described in table **5.2.1** and a typical chromatogram is shown by figure **A2.1** in appendix **A2**.

Table 5.2.1: Conditions for quantitative analysis.

Items	Conditions
Solid Phase	7.8 mm x 300 mm
Mobile Phase	1% H_2SO_4
Mobile Phase flow rate	0.5 ml/min
Analytic time	25 min
Detector	Refractive Index (RI)
Temperature	60°C
Pressure	30 bar
Sample dilution	1:10
Injection Volume	1 µl
Split Ratio	1:10

5.2.2 UV-Vis Spectroscopy

A Specord 210 UV-visible spectrophotometer from Analytik Jena (Jena, Germany) was used especially to observe the formation of nanoparticles. The UV-vis spectra were acquired at wavelengths between 200 and 800 nm. The samples were located in a quartz cell. Distilled water was used as the reference for baseline subtraction.

5.2.3 Transmission electron microscopy (TEM)

A CM200 transmission electron microscope from Philips (Eindhoven, Netherlands) was used to acquire the shape of nanoparticles. The analyses were conducted at an operation voltage of 200 kV. The X-ray emissions (EDS) analyses were conducted via the same equipment with a resolution of 77 eV to measure the element composition.

5.2.4 Dynamic light scattering (DLS)

A Zetasizer Nano-ZS from Malvern (Herrenberg, Germany) was used to measure the size of nanoparticles. The measurements were acquired with a laser power of 4 mW at a wavelength of 633 nm with a Scattering Angle of 173°. The samples were located in a quartz cell.

5.2.5   Scanning electron microscopy (SEM)

A H-S4500 FEG scanning electron microscope from Hitachi (Tokyo, Japan) was used to acquire the morphology of MCM-22. The topography was acquired based on secondary electron imaging mode.

5.2.6   X-ray diffraction (XRD)

Elemental compositions of the magnetic and nanoparticles were also determined by observing their X-ray diffraction patterns. The measurements were conducted via a PNAnalysis diffractometer from Philips (Eindhoven, Netherlands).

5.2.7   Inductively coupled plasma spectroscopy (ICP)

The measurement of metal trace was carried out by an Inductively Coupled Plasma Spectroscopy-Optical Emission Spectroscope (ICP-OES) with the brand Iris Interpid@ from Thermo Scientific (Waltham, USA). The samples were pretreated in a HNO_3/H_2SO_4 mixture and the measurements were carried out in the temperatures between room temperature and 250°C.

5.3   Experimental methods

5.3.1   Fixed-bed reaction

The catalytic glycerol oxidation reactions were conducted in an autoclave stirred tank reactor. The reactor is a bench top mini reactor (300 ml) series 4860 from Parr Instrument (Moline, USA) (figure **A3.2**). It was equipped with a stirrer, an external block heater and an internal heater. The catalyst can be introduced in the forms of pellets or granulates, where they were packed in a hollow cylindrical metal cage (figure **5.3.1**) inside the reactor. The performances of the catalytic glycerol oxidations were examined under different reaction conditions such as pH value, temperature, pressure and molar concentration of aqueous substrate solution. For carrying out the continuous fixed-bed system the reactor was constructed according to figure **5.3.2**.

Figure 5.3.1: Cylindrical metal cage for heterogeneous catalysts.

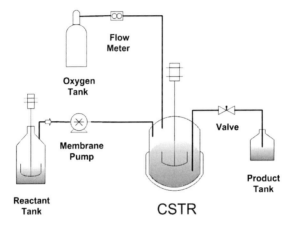

Figure 5.3.2: Flow diagram of the oxidation plant with a continuous stirred-tank reactor (CSTR).

The following procedure represents a general method for conducting the reaction: In the beginning, the catalysts were packed inside the reactor cage which then was located inside the reactor. Next, glycerol in alkaline solutions was introduced into the reactor. This was followed by raising the temperature to designated value and the pressure was adjusted. The reaction was carried out under stirring (1000 rpm). After the reaction had been finished, the reactor was cooled by immersing in an ice bath for 30 minutes.

If not otherwise mentioned, the standard reaction conditions were as follows: 10 bar air, 80°C, 1500 rpm, 8 hours of reaction time, 0.8M NaOH, 600:1 substrate to metal molar ratio, and 0.6M of glycerol initial concentration. The reaction solution was then analyzed by HPLC.

When the reactions were carried out in a continuous mode the glycerol and the gas feedings as well as products evacuation ran steady. To maintain this, the difference between the reactant introduced with the products evacuated was kept as low as possible.

5.3.2   Trickle- and fluidized bed systems

The trickle-bed reactor was constructed according to figure **5.3.3**. In the beginning, the catalysts were packed inside a fritted glass cylinder cage. The cage was then located inside a glass reactor heated with an oil bath. Oxygen and glycerol solution were introduced co-currently through the cage, where they had contact with the catalysts.

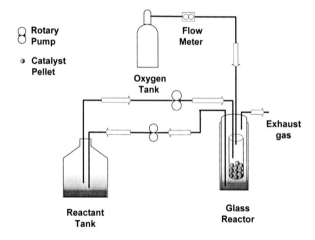

Figure 5.3.3:   A schematic diagram of a trickle-bed system.

The products and the rest of reactants departed from the fritted bottom of the cage. Then the liquid had been temporarily held inside the glass reactor before it was pumped back to the reactant tank. The gas flow rate was adjusted at 8 ml/min and the liquid flow rate was set at 15 ml/h. The standard reaction conditions were followed unless otherwise mentioned: 1 bar of oxygen, 80°C, 0.8M of NaOH, 600:1 substrate to metal molar ratio, and 0.6M of glycerol initial concentration. The reaction was carried out for 24 hours.

Similar to the trickle-bed reactor, the fluidized bed reactor was constructed according to figure **5.3.4**. The main difference to the previous system was that the catalysts were not held inside a glass cage. Therefore, they could freely move in the liquid phase inside the reactor. Another difference was the gas introduction from the lower part of the reactor resulting in a counter-current flow with the glycerol feed. The standard condition was similar to the trickle-bed system.

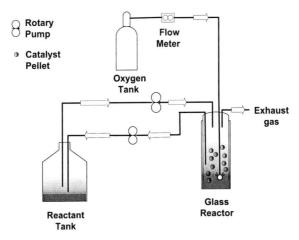

Figure 5.3.4: Schematic diagram of a fluidized bed system.

5.3.3   Synthesis of the MCM-22 precursor

The MCM-22 molecular sieve was synthesized mainly based on a procedure published by Güray and his coworkers.[137] The following example represents the general synthesis procedure to produce 8 g of the zeolite.

1.6 g of sodium hydroxide and 1.9 g of sodium aluminate were dissolved in 120 ml of distilled water. Then, this solution was added with 23.5 g of silisic acid and 12.4 g of hexamethyleneimine (HMI) as the template. The mixture was stirred and then heated for 24 hours at 45°C, then followed by another heating at 150°C for 9 days. Then the product was cooled and washed with water. Subsequently, it was dried in vacuum at 60°C overnight to produce a MCM-22 precursor. After the precursor had been calcinated at 280°C for 20 hours, 8 g of zeolite MCM-22 was produced at the end. The MCM-22 morphology was documented via the TEM analysis.

5.3.4   Synthesis of (salicylaldehyd)ethylendiamine (salen)

The method to synthesize salen molecules is as follows: 5.4 g of ethylenediamine was dissolved in 450 ml of methanol. This solution was added drop-wise with 21.9 g of salicylaldehyde then stirred for 20 minutes. Being cooled at the room temperature, salen crystals were formed. They were filtered and washed with methanol. Finally, they were dried in vacuum at 60°C overnight. A sample was analysed via an NMR instrument. With this method about 20 g of salen as crystals could be produced.

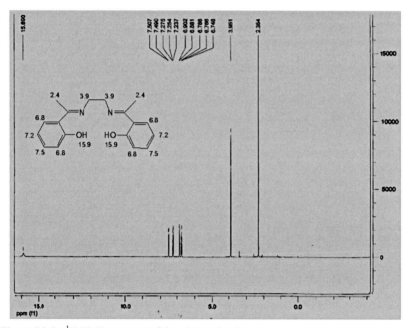

Figure 5.3.5: ^{1}H-NMR spectrum of the salen molecule.

5.3.5   Synthesis of the cobalt-salen complex

In a 100 ml twin-neck round-bottom flask equipped with a thermometer and a magnetic stirrer, 1.12 g of Co(acac)$_2$ was dissolved in 60 ml of dichloromethane. The solution was heated until boiling under nitrogen atmosphere. Then to this solution was added drop-wise 1.2 g of salen dissolved in 30 ml dichloromethane. After two hours of stirring, the mixture was cooled at room temperature. The crystal of cobalt-salen complexes were formed by adding n-pentane. Finally, they were filtered and dried in vacuum.

5.3.6   Synthesis of the cobalt-salen MCM-22 catalyst

0.1 g of cobalt-salen complex was dissolved in 40 ml of ethanol. To this solution was added 1.2 g of MCM-22, then stirred overnight at room temperature. The solid product was filtered and washed with ethanol. Subsequently, it was dried in vacuum. The dried powder was then calcinated at 280°C for 16 hours. Finally, it was washed with methanol and dried in vacuum. With this method, about 0.6 g cobalt-salen MCM-22 catalyst could be produced. The elemental composition of the samples were analysed via XRD analysis.

5.3.7   Synthesis of magnetite

The magnetite or Fe$_3$O$_4$ was synthesized mainly based on a procedure published by Xu, [138] with a few modification. The method is described as follows: 4 mmol of FeCl$_2$·4H$_2$O and 12 mmol of FeCl$_3$·6H$_2$O were dissolved in 50 ml of water and heated at 40°C. Then, 0.05 g of glycerol was added to the mixture. The solution was heated until the temperature reaches 80°C. Then, 17 ml of NH$_4$OH solution and 20 µl of 3-aminopropyltriethoxysilane (APTES) solution were added to it. The mixture was stirred for 90 min at 80°C. The reaction was taken place under inert gas atmosphere. Furthermore, the magnetite particle was separated from the original solution by applying magnetic force and washed three times with ethanol and water. Finally, it was dried under vacuum at 100°C for 6 hours.

5.3.8   Synthesis of nanoparticles

The method to prepare gold nanoparticle was based on the one reported elsewhere with few modifications[129]. The method is described as follows: 1 mmol of HAuCl$_4$·3H$_2$O was dissolved in 100 ml water and heated until boiling. Then, 10 ml of citric acid solution (40 mM) was added to the mixture and stirred for 15 min while the temperature was kept constant. Afterwards it was cooled to room temperature.

The method to prepare bimetal gold-platinum nanoparticle is described as follows: To the gold nanoparticle solution mentioned previously, 10 mmol of $H_2PtCl_6 \cdot 6H_2O$ was added. The mixture then was heated to 100°C and 1 ml of vitamin C solution (0.2 M) was added. The mixture was stirred for 5 min and cooled down to room temperature.

The method to prepare gold and gold-platinum magnetic catalysts is described as follows: To 10 ml of the gold or 5 ml gold-platinum nanoparticle solutions mentioned previously, 10 g of the magnetite particle was added. Then, the mixture was heated to 100°C and stirred for 24 hours. Finally, it was dried at 100°C under vacuum atmosphere for 6 hours.

BIBLIOGRAPHY

1. OECD.Stat. *OECD-FAO Agricultural Outlook 2012-2021: BIOFUEL - OECD-FAO Agricultural Outlook 2012-2021.* **2012**. Accessed 16 December 2012. http://stats.oecd.org/viewhtml.aspx?QueryId=36348&vh=0000&vf=0&l&il=blank&lang=en.

2. BBI International Media. *Global Biodiesel Production and Market Report.* Biodiesel Magazine **2010**. Accessed 29 September 2010. http://www.biodieselmagazine.com/article.jsp?article_id=4447.

3. Y.N. Min, F. Yan, F.Z. Liu, C.C. and, and P.W. Waldroup. *Glycerin-A New Energy Source for Poultry.* International Journal of Poultry Science. **2010**. 9(1): p. 1-4.

4. Behr, A., J. Eilting, K. Irawadi, J. Leschinski, and F. Lindner. *Improved utilisation of renewable resources: New important derivatives of glycerol.* Green Chemistry. **2008**. 10(1): p. 13-30.

5. Gallezot, P. *Selective oxidation with air on metal catalysts.* Catalysis Today. **1997**. 37(4): p. 405-418.

6. Kimura, H. *Selective Oxidation of Glycerol on a Platinum-Bismuth Catalyst by Using a Fixed-Bed Reactor.* Applied Catalysis A-General. **1993**. 105(2): p. 147-158.

7. Carrettin, S., P. McMorn, P. Johnston, K. Griffin, and G.J. Hutchings. *Selective oxidation of glycerol to glyceric acid using a gold catalyst in aqueous sodium hydroxide.* Chemical Communications. **2002**.(7): p. 696-697.

8. Christoph, R., B. Schmidt, U. Steinberner, W. Dilla, and R. Karinen. *Glycerol,* in *Ullmann's Encyclopedia of Industrial Chemistry.* **2006**. Wiley-VCH.

9. Pagliaro, M., R. Ciriminna, H. Kimura, M. Rossi, and C. Della Pina. *From glycerol to value-added products.* Angewandte Chemie-International Edition. **2007**. 46(24): p. 4434-4440.

10. BBI International Media. *Report: 12 billion gallons of biodiesel by 2020.* Biodiesel Magazine **2010**. Accessed 9 December 2010. http://www.biodieselmagazine.com/article.jsp?article_id=4145.

11. EUR-Lex. *The Directive 2003/30/EC Of The European Parliament And Of The Council Of 8 May 2003 On The Promotion Of The Use Of Biofuels Or Other Renewable Fuels For Transport.* Official Journal of the European Union. **2003**.

12. BBI International Media *Glycerin's Role in 2009*. Biodiesel Magazine. **2008**. Accessed 20 December 2008. http://www.biodieselmagazine.com/article.jsp?article_id=2976.

13. Graff, G. *Glycerin market in 2008 is a tale of two grades: Crude vs. refined*. **2008**. Purchasing.com. Accessed 9/11/2008. http://www.purchasing.com/article/222771-Glycerin_market_in_2008_is_a_tale_of_two_grades_Crude_vs_refined.php

14. China Chemical Network. *Chemical engine query glycerol offer (in Chinese)*. **2009**. Accessed 16 December 2012. http://china.chemcp.com/buy/?q=%B8%CA%D3%CD.

15. Hekmat, D., R. Bauer, and V. Neff. *Optimization of the microbial synthesis of dihydroxyacetone in a semi-continuous repeated-fed-batch process by in situ immobilization of Gluconobacter oxydans*. Process Biochemistry. **2007**. 42(1): p. 71-76.

16. Golz-Berner, K. and L. Zastrow. *Cosmetic bronzing agent based on dihydroxyacetone*. WO2005025531(A1) **2005**.

17. Gross, D. *Composition and method for treating skin*. US2009131375(A1) **2009**.

18. Rau, A.H. and H.R. Renker. *Packing materials and structures for compositions including an exothermic agent and a volatile agent*. US2007/0003675. **2007**.

19. Rahman, M.A., R. Humphreys, and S. Wu. *Biodegradable fabric conditioning molecules based on glyceric acid*. CA2151319. **1995**.

20. Tonelli, G. and J.M. Smith. *Oral compositions and method for increasing tetracycline antibiotic absorption with tartronic acid*. US3080288. **1963**.

21. Bizot, P.M., B.P. B. P. Bailey, and P.D. Hicks. *Use of tartronic acid as an oxygen scavanger*. WO9816475. **1998**.

22. Okazaki, S., K. Matoishi, and K. Itou. *Process for Producing Glycolic Acid*. US2010022740(A1) **2010**.

23. Riemenschneider, W. and M. Tanifuji. *Oxalic Acid*. **2002**. Wiley-VCH.

24. Robertson, A.J.B. *The Early History of Catalysis*. Platinum Metal Review. **1975**. **19**(2): p. 6.

25. Haines, A.H. *Relative Reactivities of Hydroxyl Groups in Carbohydrates* Advances in Carbohydrate Chemistry and Biochemistry. **1976**. **33**: p. 86-92.

26.   Porta, F. and L. Prati. *Selective oxidation of glycerol to sodium glycerate with gold-on-carbon catalyst: an insight into reaction selectivity.* Journal of Catalysis. **2004**. **224**(2): p. 397-403.

27.   Salprima, Y., R.N. Dhital, and H. Sakurai. *Gold- and gold-palladium/poly(1-vinylpyrrolidin-2-one) nanoclusters as quasi-homogeneous catalyst for aerobic oxidation of glycerol.* Tetrahedron Letters. **2011**. **52**(21): p. 2633-2637.

28.   Sankar, M., N. Dimitratos, D.W. Knight, A.F. Carley, R. Tiruvalam, C.J. Kiely, D. Thomas, and G.J. Hutchings. *Oxidation of Glycerol to Glycolate by using Supported Gold and Palladium Nanoparticles.* ChemSusChem. **2009**. **2**(12): p. 1145-1151.

29.   Kimura, H., K. Tsuto, T. Wakisaka, Y. Kazumi, and Y. Inaya. *Selective Oxidation of Glycerol on a Platinum Bismuth Catalyst.* Applied Catalysis A: General. **1993**. **96**(2): p. 217-228.

30.   Kimura, H. and K. Tsuto. *Catalytic Synthesis of Dl-Serine and Glycine from Glycerol.* Journal of the American Oil Chemists Society. **1993**. **70**(10): p. 1027-1030.

31.   Garcia, R., M. Besson, and P. Gallezot. *Chemoselective Catalytic-Oxidation of Glycerol with Air on Platinum Metals.* Applied Catalysis A: General. **1995**. **127**(1-2): p. 165-176.

32.   Prati, L. and M. Rossi. *Gold on carbon as a new catalyst for selective liquid phase oxidation of diols.* Journal of Catalysis. **1998**. **176**(2): p. 552-560.

33.   Porta, F., L. Prati, M. Rossi, S. Coluccia, and G. Martra. *Metal sols as a useful tool for heterogeneous gold catalyst preparation: reinvestigation of a liquid phase oxidation.* Catalysis Today. **2000**. **61**(1-4): p. 165-172.

34.   Demirel, S., K. Lehnert, M. Lucas, and P. Claus. *Use of renewables for the production of chemicals: Glycerol oxidation over carbon supported gold catalysts.* Applied Catalysis B: Environmental. **2007**. **70**(1-4): p. 637-643.

35.   Dimitratos, N., J.A. Lopez-Sanchez, J.M. Anthonykutty, G. Brett, A.F. Carley, R.C. Tiruvalam, A.A. Herzing, C.J. Kiely, D.W. Knight, and G.J. Hutchings. *Oxidation of glycerol using gold-palladium alloy-supported nanocrystals.* Physical Chemistry Chemical Physics. **2009**. **11**(25): p. 4952-4961.

36.   Demirel-Gülen, S., M. Lucas, and P. Claus. *Liquid phase oxidation of glycerol over carbon supported gold catalysts.* Catalysis Today. **2005**. **102-103**: p. 166-172.

37.   Prati, L., P. Spontoni, and A. Gaiassi. *From Renewable to Fine Chemicals Through Selective Oxidation: The Case of Glycerol.* Topics in Catalysis. **2009**. **52**(3): p. 288-296.

38.   Zope, B.N., D.D. Hibbitts, M. Neurock, and R.J. Davis. *Reactivity of the Gold/Water Interface During Selective Oxidation Catalysis.* Science. **2010**. **330**(6000): p. 74-78.

39.   Liang, D., J. Gao, J. Wang, P. Chen, Z. Hou, and X. Zheng. *Selective oxidation of glycerol in a base-free aqueous solution over different sized Pt catalysts.* Catalysis Communications. **2009**. **10**(12): p. 1586-1590.

40.   Brett, G.L., Q. He, C. Hammond, P.J. Miedziak, N. Dimitratos, M. Sankar, A.A. Herzing, M. Conte, J.A. Lopez-Sanchez, C.J. Kiely, D.W. Knight, S.H. Taylor, and G.J. Hutchings. *Selective Oxidation of Glycerol by Highly Active Bimetallic Catalysts at Ambient Temperature under Base-Free Conditions.* Angewandte Chemie. **2011**. **123**(43): p. 10318-10321.

41.   Liang, D., J. Gao, J.H. Wang, P. Chen, Y.F. Wei, and Z.Y. Hou. *Bimetallic Pt-Cu catalysts for glycerol oxidation with oxygen in a base-free aqueous solution.* Catalysis Communications. **2011**. **12**(12): p. 1059-1062.

42.   Takagaki, A., A. Tsuji, S. Nishimura, and K. Ebitani. *Genesis of Catalytically Active Gold Nanoparticles Supported on Hydrotalcite for Base-free Selective Oxidation of Glycerol in Water with Molecular Oxygen.* Chemistry Letters. **2011**. **40**(2): p. 150-152.

43.   Carrettin, S., P. McMorn, P. Johnston, K. Griffin, C.J. Kiely, and G.J. Hutchings. *Oxidation of glycerol using supported Pt, Pd and Au catalysts.* Physical Chemistry Chemical Physics. **2003**. **5**(6): p. 1329-1336.

44.   Bianchi, C.L., P. Canton, N. Dimitratos, F. Porta, and L. Prati. *Selective oxidation of glycerol with oxygen using mono and bimetallic catalysts based on Au, Pd and Pt metals.* Catalysis Today. **2005**. **102-103**: p. 203-212.

45.   Rodrigues, E.G., S.A.C. Carabineiro, X.W. Chen, J.A.J. Delgado, J.L. Figueiredo, M.F.R. Pereira, and J.J.M. Orfao. *Selective Oxidation of Glycerol Catalyzed by Rh/Activated Carbon: Importance of Support Surface Chemistry.* Catalysis Letters. **2011**. **141**(3): p. 420-431.

46.   Enache, D.I., J.K. Edwards, P. Landon, B. Solsona-Espriu, A.F. Carley, A.A. Herzing, M. Watanabe, C.J. Kiely, D.W. Knight, and G.J. Hutchings. *Solvent-free oxidation of primary alcohols to aldehydes using Au-Pd/TiO$_2$ catalysts.* Science. **2006**. **311**(5759): p. 362-365.

47. Dimitratos, N., C. Messi, F. Porta, L. Prati, and A. Villa. *Investigation on the behaviour of Pt(0)/carbon and Pt(0),Au(0)/carbon catalysts employed in the oxidation of glycerol with molecular oxygen in water.* Journal of Molecular Catalysis A: Chemical. **2006**. **256**(1-2): p. 21-28.

48. Ketchie, W.C., M. Murayama, and R.J. Davis. *Selective oxidation of glycerol over carbon-supported AuPd catalysts.* Journal of Catalysis. **2007**. **250**(2): p. 264-273.

49. Wang, D., A. Villa, F. Porta, D.S. Su, and L. Prati. *Single-phase bimetallic system for the selective oxidation of glycerol to glycerate.* Chemical Communications. **2006**.(18): p. 1956-1958.

50. Carrettin, S., P. McMorn, P. Johnston, K. Griffin, C.J. Kiely, G.A. Attard, and G.J. Hutchings. *Oxidation of glycerol using supported gold catalysts.* Topics in Catalysis. **2004**. **27**(1-4): p. 131-136.

51. Ketchie, W.C., Y.L. Fang, M.S. Wong, M. Murayama, and R.J. Davis. *Influence of gold particle size on the aqueous-phase oxidation of carbon monoxide and glycerol.* Journal of Catalysis. **2007**. **250**(1): p. 94-101.

52. Dimitratos, N., J.A. Lopez-Sanchez, D. Lennon, F. Porta, L. Prati, and A. Villa. *Effect of particle size on monometallic and bimetallic (Au,Pd)/C on the liquid phase oxidation of glycerol.* Catalysis Letters. **2006**. **108**(3-4): p. 147-153.

53. Dimitratos, N., F. Porta, and L. Prati. *Au, Pd (mono and bimetallic) catalysts supported on graphite using the immobilisation method: Synthesis and catalytic testing for liquid phase oxidation of glycerol.* Applied Catalysis A: General. **2005**. **291**(1-2): p. 210-214.

54. Dimitratos, N., A. Villa, C.L. Bianchi, L. Prati, and M. Makkee. *Gold on titania: Effect of preparation method in the liquid phase oxidation.* Applied Catalysis A: General. **2006**. **311**: p. 185-192.

55. Reetz, M.T. *Size-selectives Synthesis of Nanostructured Metal and Metal Oxide Colloids and Their Use as Catalysts*, in *Nanoparticles and Catalysis*, D. Astruc, Editor. **2008**. Wiley-VCH: Weinheim. p. 253-273.

56. Esumi, K., M. Shiratori, H. Ishizuka, T. Tano, K. Torigoe, and K. Meguro. *Preparation of bimetallic palladium-platinum colloids in organic solvent by solvent extraction-reduction.* Langmuir. **1991**. **7**(3): p. 457-459.

57. Demirel, S., P. Kern, M. Lucas, and P. Claus. *Oxidation of mono- and polyalcohols with gold: Comparison of carbon and ceria supported catalysts.* Catalysis Today. **2007**. **122**(3-4): p. 292-300.

58.   Huang, Z., F. Li, B. Chen, F. Xue, Y. Yuan, G. Chen, and G. Yuan. *Efficient and recyclable catalysts for selective oxidation of polyols in H2O with molecular oxygen.* Green Chemistry. **2011**. **13**(12).

59.   Villa, A., D. Wang, D.S. Su, and L. Prati. *Gold Sols as Catalysts for Glycerol Oxidation: The Role of Stabilizer.* ChemCatChem. **2009**. **1**(4): p. 510-514.

60.   Fordham, P., M. Besson, and P. Gallezot. *Selective catalytic oxidation of glyceric acid to tartronic and hydroxypyruvic acids.* Applied Catalysis A: General. **1995**. **133**(2): p. L179-L184.

61.   Abbadi, A. and H. van Bekkum. *Selective chemo-catalytic routes for the preparation of [beta]-hydroxypyruvic acid.* Applied Catalysis A: General. **1996**. **148**(1): p. 113-122.

62.   Behr, A. *Angewandte homogene Katalyse.* **2008**. Weinheim: Wiley-VCH.

63.   de Vos, D.E., B.F. Sels, and P.A. Jacobs. *Immobilization of homogeneous oxidation catalysts*, in *Advances in Catalysis*, B.C. Gates and H. Knözinger, Editors. **2001**. Academic Press: New York. p. 1-87.

64.   Baker, R.T., Sh, umacr, Kobayashi, and W. Leitner. *Divide et Impera - Multiphase, Green Solvent and Immobilization Strategies for Molecular Catalysis.* Advanced Synthesis & Catalysis. **2006**. **348**(12-13): p. 1337-1340.

65.   Sablong, R., U. Schlotterbeck, D. Vogt, and S. Mecking. *Catalysis with Soluble Hybrids of Highly Branched Macromolecules with Palladium Nanoparticles in a Continuously Operated Membrane Reactor.* Advanced Synthesis & Catalysis. **2003**. **345**(3): p. 333-336.

66.   Cole-Hamilton, D.J. *Homogeneous Catalysis--New Approaches to Catalyst Separation, Recovery, and Recycling.* Science. **2003**. **299**(5613): p. 1702-1706.

67.   Hanson, J. *Synthesis and Use of Jacobsen's Catalyst: Enantioselective Epoxidation in the Introductory Organic Laboratory.* Journal of Chemical Education. **2001**. **78**(9): p. 1266.

68.   End, N. and K.U. Schoning. *Immobilized catalysts in industrial research and application.* Immobilized Catalysts. **2004**. **242**: p. 241-271.

69.   Averseng, F., M. Vennat, and M. Che. *Grafting and Anchoring of Transition Metal Complexes to Inorganic Oxides*, in *Handbook of Heterogeneous Catalysis*, G. Ertl, et al., Editors. **2008**. Wiley-VCH: Weinheim. p. 522-538.

70. Ralph, J. *Montmorillonite*. **2010**. Mindat.org. Accessed 18 February 2010. http://www.mindat.org/min-2821.html

71. Rode, C.V., V.S. Kshirsagar, J.M. Nadgeri, and K.R. Patil. *Cobalt-salen Intercalated Montmorillonite Catalyst for Air Oxidation of p-Cresol under Mild Conditions.* Industrial & Engineering Chemistry Research. **2007**. **46**(25): p. 8413-8419.

72. Silva, A.R., M.M.A. Freitas, C. Freire, B. de Castro, and J.L. Figueiredo. *Heterogenization of a Functionalized Copper(II) Schiff Base Complex by Direct Immobilization onto an Oxidized Activated Carbon.* Langmuir. **2002**. **18**(21): p. 8017-8024.

73. Canali, L., E. Cowan, C.L. Gibson, D.C. Sherrington , and H. Deleuze. *Remarkable matrix effect in polymer-supported Jacobsens alkene epoxidation catalysts.* Chemical Communications. **1998**. p. 2561 - 2562.

74. Choudhary, D., S. Paul, R. Gupta, and J.H. Clark. *Catalytic properties of several palladium complexes covalently anchored onto silica for the aerobic oxidation of alcohols.* Green Chemistry. **2006**. **8**(5): p. 479-482.

75. Gbery, G., A. Zsigmond, and K.J. Balkus. *Enantioselective epoxidations catalyzed by zeolite MCM-22 encapsulated Jacobsen's catalyst.* Catalysis Letters. **2001**. **74**(1-2): p. 77-80.

76. Fahlman, B.D. *Materials Chemistry*. **2007**. Netherlands: Springer

77. Suramanee, P., S. Poompradub, R. Rojanathanes, and P. Thamyongkit. *Effects of Reaction Parameters in Catalysis of Glycerol Oxidation by Citrate-Stabilized Gold Nanoparticles.* Catalysis Letters. **2011**. **141**(11): p. 1677-1684.

78. Astruc, D., F. Lu, and J.R. Aranzaes. *Nanoparticles as Recyclable Catalysts: The Frontier between Homogeneous and Heterogeneous Catalysis.* Angewandte Chemie International Edition. **2005**. **44**(48): p. 7852-7872.

79. Blosi, M., S. Albonetti, M. Dondi, C. Martelli, and G. Baldi. *Microwave-assisted polyol synthesis of Cu nanoparticles.* Journal of Nanoparticle Research. **2011**. **13**(1): p. 127-138.

80. Miyamura, H., R. Matsubara, and S. Kobayashi. *Gold-platinum bimetallic clusters for aerobic oxidation of alcohols under ambient conditions.* Chemical Communications. **2008**.(17): p. 2031-2033.

81.   Wilson, O.M., M.R. Knecht, J.C. Garcia-Martinez, and R.M. Crooks. *Effect of Pd Nanoparticle Size on the Catalytic Hydrogenation of Allyl Alcohol.* Journal of the American Chemical Society. **2006. 128**(14): p. 4510-4511.

82.   Weir, M.G., M.R. Knecht, A.I. Frenkel, and R.M. Crooks. *Structural Analysis of PdAu Dendrimer-Encapsulated Bimetallic Nanoparticles.* Langmuir. **2009. 26**(2): p. 1137-1146.

83.   Endo, T., T. Yoshimura, and K. Esumi. *Synthesis and catalytic activity of gold-silver binary nanoparticles stabilized by PAMAM dendrimer.* Journal of Colloid and Interface Science. **2005. 286**(2): p. 602-609.

84.   Burato, C., P. Centomo, G. Pace, M. Favaro, L. Prati, and B. Corain. *Generation of size-controlled palladium(0) and gold(0) nanoclusters inside the nanoporous domains of gel-type functional resins: Part II: Prospects for oxidation catalysis in the liquid phase.* Journal of Molecular Catalysis A: Chemical. **2005. 238**(1-2): p. 26-34.

85.   Toshima, N. and T. Yonezawa. *Bimetallic nanoparticles—novel materials for chemical and physical applications.* New Journal of Chemistry. **1998. 22**: p. 1179 - 1201.

86.   Schmid, G. *Metal clusters and cluster metals.* Polyhedron. **1988. 7**(22-23): p. 2321-2329.

87.   Jansat, S., M. Gomez, K. Philippot, G. Muller, E. Guiu, C. Claver, S. Castillon, and B. Chaudret. *A Case for Enantioselective Allylic Alkylation Catalyzed by Palladium Nanoparticles.* Journal of the American Chemical Society. **2004. 126**(6): p. 1592-1593.

88.   Juan M. Campelo, D.L.R.L.José M.M.Antonio A.R. *Sustainable Preparation of Supported Metal Nanoparticles and Their Applications in Catalysis.* ChemSusChem. **2009. 2**(1): p. 18-45.

89.   Moshfegh, A.Z. *Nanoparticle catalysts.* Journal of Physics D: Applied Physics. **2009.**(23): p. 233001.

90.   Landau, M.V.G. *Sol-Gel Process*, in *Handbook of Heterogeneous Catalysis*, G. Ertl, et al., Editors. **2008**. Wiley-VCH: Weinhein. p. 119-156.

91.   Cheng, C., F. Xu, and H. Gu. *Facile synthesis and morphology evolution of magnetic iron oxide nanoparticles in different polyol processes.* New Journal of Chemistry. **2011. 35**(5): p. 1072–1079.

92. Teranishi, T. and N. Toshima. *Preparation, Characterization, and Properties of Bimetallic Nanoparticles*, in *Catalysis and Electrocatalysis at Nanoparticle Surfaces*, A. Wieckowski, E.R. Savinova, and C.G. Vayenas, Editors. **2003**. CRC Press. p. 379-403.

93. Toneguzzo, P., G. Viau, O. Acher, F. Guillet, E. Bruneton, F. Fievet-Vincent, and F. Fievet. *CoNi and FeCoNi fine particles prepared by the polyol process: Physicochemical characterization and dynamic magnetic properties.* Journal of Materials Science. **2000**. 35(15): p. 3767-3784.

94. Liu, C., X.W. Wu, T. Klemmer, N. Shukla, X.M. Yang, D. Weller, A.G. Roy, M. Tanase, and D. Laughlin. *Polyol process synthesis of monodispersed FePt nanoparticles.* Journal of Physical Chemistry B. **2004**. 108(20): p. 6121-6123.

95. Beck, W., C.G.S. Souza, T.L. Silva, M. Jafelicci, and L.C. Varanda. *Formation Mechanism via a Heterocoagulation Approach of FePt Nanoparticles Using the Modified Polyol Process.* The Journal of Physical Chemistry C. **2011**. 115(21): p. 10475-10482.

96. Toneguzzo, P., O. Acher, G. Viau, A. Pierrard, F. Fievet-Vincent, F. Fievet, and I. Rosenman. *Static and dynamic magnetic properties of fine CoNi and FeCoNi particles synthesized by the polyol process.* Ieee Transactions on Magnetics. **1999**. 35(5): p. 3469-3471.

97. Fievet, F., F. Fievetvincent, J.P. Lagier, B. Dumont, and M. Figlarz. *Controlled Nucleation and Growth of Micrometer-Size Copper Particles Prepared by the Polyol Process.* Journal of Materials Chemistry. **1993**. 3(6): p. 627-632.

98. Das, M., P. Dhak, S. Gupta, D. Mishra, T.K. Maiti, A. Basak, and P. Pramanik. *Highly biocompatible and water-dispersible, amine functionalized magnetite nanoparticles, prepared by a low temperature, air-assisted polyol process: a new platform for bio-separation and diagnostics.* Nanotechnology. **2010**. 21(12): p. 1-12.

99. Jalem, R., R. Koike, Y. Yang, M. Nakayama, and M. Nogami. *Electrochemical characterization of a porous Pt nanoparticle "Nanocube-Mosaic" prepared by a modified polyol method with HCl addition.* Nano Research. **2011**. 4(8): p. 746-758.

100. Nishioka, M., M. Miyakawa, Y. Daino, H. Kataoka, H. Koda, K. Sato, and T.M. Suzuki. *Rapid and Continuous Polyol Process for Platinum Nanoparticle Synthesis Using a Single-mode Microwave Flow Reactor.* Chemistry Letters. **2011**. 40(12): p. 1327-1329.

101. Biacchi, A.J. and R.E. Schaak. *The Solvent Matters: Kinetic versus Thermodynamic Shape Control in the Polyol Synthesis of Rhodium Nanoparticles.* ACS Nano. **2011.** **5**(10): p. 8089-8099.

102. Wiley, B., T. Herricks, Y.G. Sun, and Y.N. Xia. *Polyol synthesis of silver nanoparticles: Use of chloride and oxygen to promote the formation of single-crystal, truncated cubes and tetrahedrons.* Nano Letters. **2004.** **4**(9): p. 1733-1739.

103. Geus, J.W. and A.J. van Dillen. *Preparation of Supported Catalysts by Deposition-Precipitation,* in *Handbook of Heterogeneous Catalysis,* G. Ertl, et al., Editors. **2008.** Wiley-VCH: Weinheim. p. 428-465.

104. Hughes, M.D., Y.J. Xu, P. Jenkins, P. McMorn, P. Landon, D.I. Enache, A.F. Carley, G.A. Attard, G.J. Hutchings, F. King, E.H. Stitt, P. Johnston, K. Griffin, and C.J. Kiely. *Tunable gold catalysts for selective hydrocarbon oxidation under mild conditions.* Nature. **2005.** **437**(7062): p. 1132-1135.

105. Bronkala, W.J. *Magnetic Separation,* in *Ullmann's Encyclopedia of Industrial Chemistry.* **2000.** Wiley-VCH: Weinheim.

106. Iacob, G., A.D. Ciochina, and O. Bredetean. *High Gradient Magnetic Separation Ordered Matrices.* European Cells and Materials. **2002.** **3**: p. 167-169.

107. Gubin, S.P. *Introduction,* in *Magnetic Nanoparticle,* S.P. Gubin, Editor. **2009.** Wiley-VCH. p. 1-23.

108. Kolesnichenko, V.L. *Synthesis of Mamoparticulate Magnetic Materials,* in *Magnetic Nanoparticle,* S.P. Gubin, Editor. **2009.** Wiley-VCH: Weinheim. p. 25-58.

109. Aschwanden, L., P. Barbara, R. Peggy, K. Beat, and B. Alfons. *Magnetically Separable Gold Catalyst for the Aerobic Oxidation of Amines.* ChemCatChem. **2009.** **1**(1): p. 111-115.

110. Polshettiwar, V. and R.S. Varma. *Nanoparticle-supported and magnetically recoverable palladium (Pd) catalyst: a selective and sustainable oxidation protocol with high turnover number.* Organic & Biomolecular Chemistry. **2009.** **7**(1): p. 37-40.

111. Tong, J.H., L.L. Bo, Z. Li, Z.Q. Lei, and C.G. Xia. *Magnetic $CoFe_2O_4$ nanocrystal: A novel and efficient heterogeneous catalyst for aerobic oxidation of cyclohexane.* Journal of Molecular Catalysis A-Chemical. **2009.** **307**(1-2): p. 58-63.

112. Rahimpour, M.R., S.M. Jokar, and Z. Jamshidnejad. *A novel slurry bubble column membrane reactor concept for Fischer–Tropsch synthesis in GTL technology.* Chemical Engineering Research and Design. **2012.** **90**(3): p. 383-396.

113.   Werther, J. *Fluidized-Bed Reactors*, in *Handbook of Heterogeneous Catalysis*, G. Ertl, et al., Editors. **2008**. Wiley-VCH: Weinheim. p. 2106-2131.

114.   Gallucci, F., M. van Sint Annaland, and J.A.M. Kuipers. *Comparison of packed bed and fluidized bed membrane reactors for methane reforming*. in *6th International Symposium on Multiphase Flow, Heat Mass Transfer and Energy Conversion*. **2009**. Xi'an, China.

115.   Rahimpour, M.R. and H. Elekaei. *Optimization of a novel combination of fixed and fluidized-bed hydrogen-permselective membrane reactors for Fischer–Tropsch synthesis in GTL technology*. Chemical Engineering Journal. **2009**. **152**(2–3): p. 543-555.

116.   Nedeltchev, S. and A. Schumpe. *Slurry Reactors*, in *Handbook of Heterogeneous Catalysis*, G. Ertl, et al., Editors. **2008**. Wiley-VCH: Weinheim. p. 2132-2156.

117.   Chaudhari, R.V. and P.A. Ramachandran. *Three phase slurry reactors*. AIChE Journal. **1980**. **26**(2): p. 177-201.

118.   Fox, J.M., B.D. Degen, G. Cady, F.D. Deslate, and R.L. Summers. *Slurry Reactor Design Studies. Slurry vs. Fixed-Bed Reactors for Fischer-Tropsch and Methanol: Final Report*. **1990**. Emerging Fuels Technology. Accessed 30 October 2012. http://www.fischer-tropsch.org/DOE/DOE_reports/91005752/de91005752_toc.htm

119.   de la Peña O'Shea, V.A., M.C. Álvarez-Galván, J.M. Campos-Martín, and J.L.G. Fierro. *Fischer–Tropsch synthesis on mono- and bimetallic Co and Fe catalysts in fixed-bed and slurry reactors*. Applied Catalysis A: General. **2007**. **326**(1): p. 65-73.

120.   Hu, W., D. Knight, B. Lowry, and A. Varma. *Selective Oxidation of Glycerol to Dihydroxyacetone over Pt−Bi/C Catalyst: Optimization of Catalyst and Reaction Conditions*. Industrial & Engineering Chemistry Research. **2010**. **49**(21): p. 10876-10882.

121.   Demirel, S., M. Lucas, J. Warna, D. Murzin, and P. Claus. *Reaction kinetics and modelling of the gold catalysed glycerol oxidation*. Topics in Catalysis. **2007**. **44**(1-2): p. 299-305.

122.   Tsutomu, M., U. Masaharu, W. Naiwei, Sh, umacr, and Kobayashi. *Recent Advances in Immobilized Metal Catalysts for Environmentally Benign Oxidation of Alcohols*. Chemistry - An Asian Journal. **2008**. **3**(2): p. 196-214.

123.   Structure Commission of the International Zeolite Association. *Framework Type MWW*. **2007**. Accessed 12 August 2012. http://izasc-mirror.la.asu.edu/fmi/xsl/IZA-

SC/ftc_image.xsl?-db=FWimages&-
lay=FWimages&FWimages=MWW_large_cage.jpg&-find.

124. Xu, G., X. Zhu, X. Niu, S. Liu, S. Xie, X. Li, and L. Xu. *One-pot synthesis of high silica MCM-22 zeolites and their performances in catalytic cracking of 1-butene to propene.* Microporous and Mesoporous Materials. **2009**. **118**(1–3): p. 44-51.

125. Varanda, L.C., M. Imaizumi, F.J. Santos, and M. Jafelicci. *Iron Oxide Versus $Fe_{55}Pt_{45}/Fe_3O_4$: Improved Magnetic Properties of Core/Shell Nanoparticles for Biomedical Applications.* Ieee Transactions on Magnetics. **2008**. **44**(11): p. 4448-4451.

126. Yang, W.-C., R. Xie, X.-Q. Pang, X.-J. Ju, and L.-Y. Chu. *Preparation and characterization of dual stimuli-responsive microcapsules with a superparamagnetic porous membrane and thermo-responsive gates.* Journal of Membrane Science. **2008**. **321**(2): p. 324-330.

127. Koneracká, M., V. Závišová, M. Timko, P. Kopčanský, N. Tomašovičová, and K. Csach. *Magnetic Properties of Encapsulated Magnetite in PLGA Nanospheres.* acta physica polonica A. **2008**. **113**: p. 4.

128. Can, K., M. Ozmen, and M. Ersoz. *Immobilization of albumin on aminosilane modified superparamagnetic magnetite nanoparticles and its characterization.* Colloids and Surfaces B: Biointerfaces. **2009**. **71**(1): p. 154-159.

129. Shaojun, G., D. Shaojun, and W. Erkang. *A General Route to Construct Diverse Multifunctional Fe_3O_4/Metal Hybrid Nanostructures.* Chemistry - A European Journal. **2009**. **15**(10): p. 2416-2424.

130. Wang, X., S. Huang, Z. Shan, and W.S. Yang. *Preparation of $Fe_3O_4@Au$ nano-composites by self-assembly technique for immobilization of glucose oxidase.* Chinese Science Bulletin. **2009**. **54**(7): p. 1176-1181.

131. ICIS. *Indicative Chemical Prices A-Z.* **2012**. Accessed 12 August 2012 http://www.icis.com/chemicals/channel-info-chemicals-a-z/.

132. Katryniok, B., H. Kimura, E. Skrzynska, J.-S. Girardon, P. Fongarland, M. Capron, R. Ducoulombier, N. Mimura, S. Paul, and F. Dumeignil. *Selective catalytic oxidation of glycerol: perspectives for high value chemicals.* Green Chemistry. **2011**. **13**(8): p. 1960-1979.

133. Reed Business Information Limited. *25th January 2012 Glycerine (US Gulf).* **2012**. Accessed 30 March 2012 http://www.icispricing.com/il_shared/Samples/SubPage170.asp.

134. Rieser, K.-P. *BASF raises prices of formic acid in Europe, Asia and South America.* BASF SE. **2011.** http://www.asiapacific.basf.com/apex/AP/AsiaPacific/en/function/conversions:/publis h/AsiaPacific/upload/Press2011/pdf/Raise_prices_formic_acid_Europe_Asia_South_ America_16Jun2011.pdf.

135. TCI America. *DL-Glyceric Acid* **2012.** Accessed 15 August 2012. http://www.tcichemicals.com/eshop/en/us/commodity/D0602/.

136. Zhang, S., Y. Shao, G. Yin, and Y. Lin. *Electrostatic Self-Assembly of a Pt-around-Au Nanocomposite with High Activity towards Formic Acid Oxidation13.* Angewandte Chemie. **2010.** **122**(12): p. 2257-2260.

137. Güray, I., J. Warzywoda, N. Baç, and A. Sacco Jr. *Synthesis of zeolite MCM-22 under rotating and static conditions.* Microporous and Mesoporous Materials. **1999.** **31**(3): p. 241-251.

138. Xu, L., M.-J. Kim, K.-D. Kim, Y.-H. Choa, and H.-T. Kim. *Surface modified Fe₃O₄ nanoparticles as a protein delivery vehicle.* Colloids and Surfaces A: Physicochemical and Engineering Aspects. **2009.** **350**(1-3): p. 8-12.

APPENDICES

A1   Definition of conversion, selectivity and yield

Conversion (X) is the ratio between the amount of glycerol which had been converted and the initial glycerol amount. This is described by equation A1.1. Selectivity (S) towards a particular substance represents the ratio between the amount of that substance produced and the amount of converted glycerol after t hours of reaction. This relation is described by equation A1.2. Finally, the yield (Y) towards a particular substance is defined as the ratio between the amount of that substance produced and the initial glycerol amount.

$$X = \frac{n_0 - n_t}{n_0} \times 100\% \qquad\qquad \text{equation A1.1}$$

$$S = \frac{n_{i,t}}{n_0 - n_t} \times \frac{|v_G|}{|v_i|} \times 100\% \qquad\qquad \text{equation A1.2}$$

$$Y = \frac{n_{i,t}}{n_0} \times \frac{|v_G|}{|v_i|} \times 100\% \qquad\qquad \text{equation A1.3}$$

Where n_0 and n_t represent the amount of glycerol before the reaction and after t hours of reaction, respectively. $n_{i,t}$ is the amount of that substance produced after t hours of reaction. v_G and v_i are the stoichiometric coefficients of glycerol and a particular substance, respectively.

A2   Spectrum

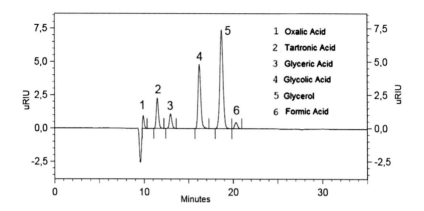

Figure A2.1:   A typical HPLC chromatogram.

Table A2.1:   The retention times of the substances in the HPLC chromatogram.

Substances	Retention time [minutes]
Oxalic Acid	10.2
Tartronic acid	11.6
Glyceric acid	12.8
Glycolic acid	16.2
Glycerol	18.7
Formic Acid	20.4

Analysis conditions are mentioned in table **5.2.1**.

A3 Magnetic slurry system

The realization of the magnetic slurry system depicted in figure **3.3.1** is displayed in figure **A3.1**. Moreover, the specifications of major instruments are listed in table **A3.1**.

Figure A3.1: The miniplant of the magnetic slurry system.

Table A3.1: List of important instruments in the miniplant displayed in figure **A3.1**.

Instrument	Producer	Model
Peristaltic Pumps	Multifix (Schwäbisch Gmünd, Germany), Ismatec (Wertheim, Germany)	M80, Reglo
Membrane Pumps	Prominent (Heidelberg, Germany)	A2001, Gamma4
Balances	Mettler Toledo (Gießen, Germany)	XS6002S, PM4600
Electromagnet	Intertec (Freising, Germany)	IST-MS 5030, 600 N

Figure A3.2: Pressure reactor.

Figure A3.3: Magnetic separation part.

A4   Solutions of nanoparticles and magnetic particles

Figure A4.1: Solutions of gold precursor (left), gold nanoparticles (middle) and platinum nanoparticles.

Figure A4.2: Appearance of magnetite or Fe_3O_4 (left) and maghemite (γ-Fe_2O_3) (right) after stability tests.

CURRICULUM VITAE

Ken A. Irawadi
Born on 4 March 1977 in Bogor, Indonesia

10/ 2006 – 12/2012	**Ph.D. Study in Process Engineering** at the Technical University of Dortmund, Germany
Since 2005	**University Lecturer** at Bogor Agricultural University
08/ 2001 – 05/ 2004	**M.Sc. Study in Process Engineering** at the Technical University of Hamburg-Harburg, Germany
2000-2001	**Assistant Lecturer** at Bogor Agricultural University
08/ 1995 – 12/ 1999	**B.Sc. Study in Agroindustrial Technology** at Bogor Agricultural University, Indonesia